the cosmic zoo...

the cosmic Z

scale,
knowledge,
and
mediation

zachary horton

The University of Chicago Press

Chicago and London

The University of Chicago Press, Chicago 60637
The University of Chicago Press, Ltd., London
© 2021 by The University of Chicago
Published 2021
Printed in the United States of America

30 29 28 27 26 25 24 23 22 21   1 2 3 4 5

ISBN-13: 978-0-226-74230-4 (cloth)
ISBN-13: 978-0-226-74244-1 (paper)
ISBN-13: 978-0-226-74258-8 (e-book)
DOI: https://doi.org/10.7208/
chicago/9780226742588.001.0001

Library of Congress Cataloging-in-Publication Data

Names: Horton, Zachary K., author.
Title: The cosmic zoom : scale, knowledge, and mediation /
    Zachary Horton.
Other titles: Scale, knowledge, and mediation
Description: Chicago ; London : The University of Chicago Press,
    2021. | Includes bibliographical references and index.
Identifiers: LCCN 2020040705 | ISBN 9780226742304 (cloth) |
    ISBN 9780226742441 (paperback) | ISBN 9780226742588
    (ebook)
Subjects: LCSH: Scale (Philosophy)
Classification: LCC B105.S33 H67 2021 | DDC 121/.34—dc23
LC record available at https://lccn.loc.gov/2020040705

♾ This paper meets the requirements of
ANSI/NISO Z39.48-1992 (Permanence of Paper).

For those who call distant regions
of the scalar spectrum home

contents

scale theory

THE RABBIT HOLE

Curious Alice falls down a hole. Hurled into a radically unfamiliar world, she must quickly adapt to its alien logic. Ingesting certain substances, she discovers, causes her to change size. This revelation provides both a challenge and an opportunity. Each shift of scale alters her perspective on and relationship to Wonderland, complicating her quest to map its terrain and logics. Yet these very shifts in perspective expand her possibilities for apprehending and interacting with the environment's strange features.

The resourceful child passes briskly through three stages of scalar awareness. In the first, she finds herself inexplicably the wrong size for the task at hand. Her scale has become a handicap: "when she got to the door, she found she had forgotten the little golden key, and when she went back to the table for it, she found she could not possibly reach it."[1] The door is a threshold that admits only those who have mastered the antechamber's multiscalar mechanism. Frame, door, lock, table, key—each belongs to a different scale. How to unite them on a single plane?

In the second stage of scalar awareness, Alice learns through experimentation that she can access other scales—eating cakes and drinking potions, she grows and shrinks, shifting her scalar relationship with her environment—but she cannot yet control her metamorphoses. In the third stage, after experimenting with the Caterpillar's mushroom, she learns to discipline her scale jumping, to contain it within narrow bounds, to tame the metamorphoses and deploy them strategically. It is in this third stage that Alice, pockets provisioned with fungi, finally unlocks the door to the garden and there encounters and nearly overcomes the pack of sentient playing cards that enforce Wonderland's laws. Through it all, Alice herself never alters more than her relative size. Her fundamental identity as an educated, upper-class Englishwoman-in-training remains invariant. She uses her abilities to master the logics of Wonderland but can't seem to understand why her anthropocentric assumptions continually upset her nonhuman interlocutors. "I wish the creatures wouldn't be so easily offended!" she thinks to herself, placing the blame squarely on their shoulders.[2] She has learned, that is, to *access* scalar alterity, but not to absorb its dynamics, to become herself other. *Alice in Wonderland* is a fable about scalar alterity, about how one scale comes to dominate another.

The Cosmic Zoom argues that as a species we are in a situation analogous to that of Alice. We have, like her, passed through three stages, from *awareness* of scalar alterity to blundering *encounter* to disciplined *access*. Like Alice, we injure the other beings we encounter through our anthropocentric attempts to force them to conform to our rules—rules developed by us, for us, at our native scale. We have extended the scales of our knowledge by disciplining it and are continually prying open new scales and forcing them to conform to the logics of our institutions and ideologies. As we unlock the doors of Wonderland, each leading to a new scalar milieu, we are apt to play out our story according to the well-worn narrative logics of colonization and extraction that Western culture has long rehearsed in its media, even knowing that on this path only environmental apocalypse awaits.

This is a book about scale. Its central purpose is to outline a genealogy of the concept and to build a critical transdisciplinary vocabulary and framework that will enable a larger dialogue. Such dialogue has until now progressed only in fits and starts, hampered by incommensurate theoretical frameworks and practical methodologies in different domains of knowledge. The most common way to approach scale in the humanities, for instance, is to assume a scalar axis, from the small to the large, and track how an artwork, text, technology, event, or discourse has engaged or produced a larger (or occasionally, smaller) scale than previously: a larger geographical milieu, a wider social scope, a greater word count, a national or imperial or local sensibility, a greater audience, an interest in bacteriological and viral domains, and so on. This is a scholarship of scalar *access*. I want to differentiate that approach to scale from the central concerns of the book you are reading.

My goal in this volume is not to rehearse the scalar march of human cultural production to ever expanded scales along a small-large axis but to engage an entirely different axis: one of scalar alterity, which runs from the "pole" of *scalar difference* to that of *scalar collapse*, or the speculative conjoining of different scales within a single medium, eliding the qualitative differences between them. To investigate not individual scales but scalar dynamics themselves, I propose that we dig deeper in order to understand how and in what way our received size-domain axis emerges and what cultural work it performs.

While this study does engage questions of scalar access, its operative question is not *what* is accessed (a social group, a class, a city, a region, an empire, a nation, the planet, a genome, the cosmos) but rather *how* access is mediated in ways that engage or occlude scalar difference. The genealogy of scale that emerges is historically discontinuous and disarticulated from progressive expansions of access, focused rather on moments in which scalar mediation itself innovates techniques that frame our scalar access in new ways, stabilizing the protocols of such access either to enable the exploitation of other scales or to open up new forms of encounter. The central focus of this book is not, that is, the human experience of scale effects but, rather, the radically nonhuman dynamics and potentials of scale, as a concept and as a form of mediation. Engaging scalar alterity, we will see, implies a dismantling of the edifices of humanism itself, or at least the specialized mode that I call *pan-scalar humanism*.

Pan-scalar humanism is a tradition that tames the alterity of different scales by relativizing it, binding unfamiliar scales to the familiar ones of the human. It arose out of an Enlightenment notion of the absolute value, autonomy, and centrality of the human subject that, *a priori*, colors all potential trans-scalar encounters. Pan-scalar humanism frames all trans-scalar encounters as either extensions of the human into analogous scales (collapsing scalar difference) or as the beneficent extension of the human lifeworld into frontier scales. The human thus becomes a scalar technique of assimilation and colonization applicable in theory to all scales of reality. At the same time, pan-scalar humanism mobilizes radically disparate scales to buttress the human subject. *Alice in Wonderland* satirizes pan-scalar humanism without yet being able to articulate an alternative scalar economy or politics.

4

In order to examine the relationship between the human, mediation, and scale, I have chosen to focus on the "cosmic zoom," a self-consciously medial project that attempts to characterize the scalar articulations of the cosmos by visualizing, from a single perspective, a spectrum of scales from the largest to the smallest known.[3] The cosmic zoom has taken textual, imagistic, motion picture, and new media forms. The most famous instantiation is *Powers of Ten*, a 1977 film by designers Ray and Charles Eames that begins with two picnickers in a field, zooms out to encompass the entire universe, then zooms in again until the nucleus of a single carbon atom fills the frame. I discuss this film at length in chapter 4, but as we shall see in the antecedent chapters, significant examples of the cosmic zoom preceded it, beginning with the texts discussed in this chapter and emerging in fully modern form in the 1950s with *Cosmic View*, an influential book by a radical Dutch educator.

Using Michel Foucault's term for the unacknowledged framework that grounds the "conditions of possibility" of knowledge itself,[4] we might call the cosmic zoom the perfect encapsulation of a scalar *épistémè*, a set of scales that have been stabilized as legible environments and therefore objects of knowledge. The politics of the cosmic zoom are the politics of a culture's engagement with scale, dredged up from their subterranean depths and gilded for human consumption. Individual instantiations of the cosmic zoom can be imaginative or conservative, radical or reactionary. Developing a critical vocabulary and analytic framework for the cosmic zoom, then, is tantamount to developing a theory of scale itself.

The theory of scale that I develop in this book has, at its core, a simple premise: scale is a primary form of difference. It is primary in the sense that it is present on the scene and does its work before stable identities (subjects or objects) have formed along its spectrum. But it is also caught up in the operations of thought as entities begin to navigate their environments, stabilizing them into scaled milieus. Scale, as a series of relational dynamics, is thus a circuit: an irruption of the new on one side, and its ordering for others on the obverse. These dual scalar processes must be modulated in order to form a legible, navigable plane. Thus media, or rather mediation (as process), are fundamental to the operations of this scalar circuit. How scale mediates between an observing entity and the details of its environment is just as important, in what follows, as how conceptual, narrative, and technological mediation produce and stabilize individual scales.

While often conflated with size, scale has many facets and is difficult to define. It names a set of relations: external relations between two or more milieus, and internal relations between entities within a single milieu. Scale is both a stabilizing process by which particular milieus (scales) emerge as defined domains of (inter)action and the differential potentials that arise between such stabilized milieus. Scale thus implies both a politics and an environment.

HUMANITY IN WONDERLAND

The world has changed. It may be difficult to pinpoint the exact moment we fell down the rabbit hole, but we know that we are now in Wonderland. Scales that we, as humans, had taken for granted are suddenly front and center, blocking the path both forward and back. And when we look around, everything seems different, as if the borders between things have slipped out of focus. We may adjust our glasses and wipe our screens, but it isn't entirely clear whether it is the "human" or the "environment" that has changed scale. There are moments, such as during the COVID-19 pandemic, that crystallize Wonderland's weirdly scrambled scales: global ligaments become viral infrastructure, while unmediated contact between human beings comes to seem grotesquely inhuman. These temporary crystallizations, so easily narrativized as anomalies, actually reveal long-term, profound scalar shifts in our technocultural milieus. The old-fashioned, comforting human scale has crumbled, and our species now stands poised between global climate change and big data. The latter has recomposed the human along new scalar fault lines, eliding the meso-scale individual as such while producing on-the-fly collectives from micro and macro attributes that humans can experience only through highly mediated means. Meanwhile, we are staring into the abyss of the most significant threat that the human and many nonhuman inhabitants of Earth have faced during our tenure as the planet's keystone species: global climate change.

Big data and anthropogenic climate change are the two sides of Wonderland, united through and as new scalar dynamics. Human actions that we conceive of as individual choices, governmental policies promoting national competitiveness and self-regulation, and trillions of economic exchanges predicated upon a rationalist, individualist logic no longer seem natural when viewed in the aggregate, at larger scales. The cumulative effect of billions of such "everyday" exchanges, climate change is a prime example of *scalar magic*: when the attention of any observer is fixed at one scale only, scaling phenomena will seem to vanish into thin air—when all they have really done is shift to other scales. This describes a scalar logic that is at once materially causal, ethically laden, and politically tactical.

What is scale? We could do worse than to begin with these three dimensions: Scale is, first, an *ontological determinant* in that it dictates how certain physical states become other physical states. Growth, depletion, division, aggregation—these are shifts between scales, not merely metamorphoses of individual forms. Scale is, second, an *ethical ground* that binds individuals, groups, and territories into interconnected milieus of interdependence and responsibility. And scale is, third, a set of *political tactics* for aggregating and disaggregating assemblages. In this final capacity, contemporary scalar politics invests energy into singularities (individual heroes and villains, monuments, memes) and thus away from systems, while displacing undesirable consequences to nonvisible scales: the vast ocean, the atmosphere, the nano realm, the far future—comfortingly distant points on the scalar spectrum. But these scalar deferrals always return eventually. Rob Nixon refers to damaging processes that have been kept out of sight by assigning them to other scales as "slow violence." The damage is done to environments, humans, and communities, but it is "of indifferent interest to the sensation-driven technologies of our image-world," which require that events be concentrated in space and time to become visible as such.[5] Slow violence is not accidental; it is the result of weaponized scalar difference. How trans-scalar flows are managed and put to work, and what they produce for whom, are questions any contemporary analytic of power must ask.

A newfound scale mania has infected even those not professionally interested in questions of metamorphosis or power. Everyone seems to sense that they are in Wonderland, a new trans-scalar environment. Scale is on everyone's lips. Films that feature shrinking and expanding humans, such as Alexander Payne's *Downsizing* (2017) and Marvel's *Ant-Man* franchise, are enjoying a resurgence in popularity.[6] In the past decade, the longstanding preoccupation with scale in the natural sciences has diffused into many fields within the humanities and social sciences, concomitant with a surge of interest in the subject in popular science literature, including recent coffee-table volumes such as Hooft and Vandoren's *Time in Powers of Ten* and Caleb Scharf's *The Zoomable Universe*. We frequently speak of the scale of big data, of climate disasters, of drought, of pandemic, of rainforest destruction, of radioactive contamination, of airborne and waterborne pollution, of economic recession, of the mass displacements of refugees, of arctic and antarctic ice sheet depletion, of the sixth mass extinction. These otherwise disparate discourses share a common thread of scalar enthusiasm and scalar dread. Both are reflexive: their common narrative trope is the sounding of the scale of the human. Efforts to accomplish this have tended to posit a "species being," characterizing the human at a new temporal and spatial scale—the global, the planetary, or the geological. This approach provides some critical distance from the human individual as the alpha and omega of meaning and value; it is nonetheless subject to the same panoply of essentialisms, apologetics, and homogenizations. As Derek Woods asks, "Is the concept of the human scalable? To answer this question, we need scale critique to grasp what 'human' means when it names the subject of the Anthropocene."[7] My approach in this book is to take the human as a multiscalar flux inextricably caught up in a flowing network of scales, sometimes as a fragment in larger bodies and sometimes as an environment for smaller ones. The human is scale-unstable, even as human media infrastructures and disciplinary knowledge practices seek to stabilize particular scales.

One of the consequences of taking up the question of "the human" from the perspective of scale is that it ceases to function as a boundary or membrane (conceptual or corporeal) between an inside and an outside. This is partly because the human is "trans-corporeal" in Stacy Alaimo's sense of uncontainably embodied and "always inter-meshed with the more-than-human world."[8] But further, human knowledge production has always, first and foremost, proceeded from a taming of scale. Containing scalar difference within stabilized domains and organizing those domains into a spatially and conceptually continuous plane anchored by the unmarked scale of the human is central to the project of humanism. Indeed, these mediations inaugurate the emergence of a scale-stable human subject in the first place. To reverse our analytical priority, to begin with scale itself rather than to enumerate it as an attribute of an already unified subject or object, profoundly disturbs this scalar pact of humanist thought. It may well be that humanism is ill-suited as a response to the challenges of Wonderland. At the very least we will need to change our default question: instead of asking which scales are occupied or accessed by a conceptually preconstituted human, we'll need to ask how the human emerges—along with many other objects and subjects—out of the dynamics of scale.

Welcome to Wonderland's feedback loop: the human emerges from fundamental scalar differentiation (explored in detail in chapter 5), stabilizes certain scales through discursive and medial infrastructures, and then harnesses them for further production. The result is a kind of "scalar accumulation," strata upon strata of produced objects and subjects organized and sorted according to their naturalized scales. Whether we analyze this in the form of capitalism, climate change, or big data, the result is the same: ever increasing scales of accumulation take the form of a widening milieu of the human, organized concentrically around our native scale—that is, the scale of our immediate sensory field. In economic terms, this means continually expanding markets into new regions and temporalities, what David Harvey refers to as "spatio-temporal fixes" for capital's overaccumulation.[9] Now, however, this expansion is not merely geographical and cultural, but also trans-scalar: neoliberal capital is increasingly exploiting new scalar frontiers, from the solar system to future temporalities to the fabled "radical abundance" of the nanoscale.[10]

This dynamic is not, of course, sustainable: scalar alterity is not a matter of linear differentials (more or less of something, such as capital) but rather of radical discontinuities in the scalar spectrum. Any system predicated upon the continual appropriation and stabilization of new scales in the service of a single master scale (as a dominant and homogenizing logic) is bound to run up against its absolute limits relatively quickly, whether those limits take the form of a financial crash, a global pandemic, a massive loss of biodiversity, the tipping point of global climate dynamics, technological singularity, or the structural collapse of human civilization.

It may seem as though the very forays into other scales that brought us to Wonderland will necessarily prove our undoing. Yet this is not merely a question of having opened Pandora's box. Our encounters with other scales may be dangerous but also open up the possibility of our being remade *at* other scales. We believe we construct scale, but "our" scalar mediation confronts us with entities as terrifying and wondrous as supernovas and nuclear fission, sea-level rise and computer viruses, galactic spirals and quantum uncertainty. New forms of subjectivity are continually produced by these trans-scalar encounters. By "trans-scalar encounters" I mean the catalyzing events that take place when an observer adapted to a milieu defined by a particular scale of typical events encounters structures and processes at a different scale. The trans-scalar encounter is an encounter with difference and can therefore be either generative of further differentiation or a form of colonial capture, the imprinting of the dynamics of a socially engineered human scale onto another. Unfortunately, most of this occurs without any self-reflexive register in the realm of thought itself. Like Alice, we blunder into trans-scalar encounter without even knowing the local customs.

If the human is not the protagonist of the trans-scalar encounter, it becomes one subjectivizing effect among others. At issue here is how one region of the scalar spectrum comes to encounter another, discontinuous region. Understanding scale as a processual differentiation through encounter helps to challenge the unidirectional model of observer and observed. All of existence involves continuous trans-scalar encounter, but the discontinuity inherent in the process is reciprocal: each scale is stabilized only through encounter, while encounters always begin from a particular scale. There is no difference between the observer and the observed—perspectives can emerge from any point on the scalar spectrum, along with subjectivities to inhabit them. Media theory furnishes us with important tools to help us theorize this destabilization of the relationship between observer and observed, which I address most directly in chapters 3 and 4 in relation to Ray and Charles Eames's cosmic-zoom films.

These, then, are the problems of Wonderland: How to think larger and smaller than the human scale? How to think with the nonhuman? How to incorporate a multiscalar form of thought without homogenizing detail and difference? My goal is to frame these problems and suggest the beginnings of solutions through an analysis of Wonderland's most scale-reflexive medial form: the cosmic zoom.

THE COSMIC ZOOM

This book develops a theory of scale as primary difference at the same time that it works to grow new connective tissues between our understandings of mediation, scale, and subjectivity. It is a necessarily experimental project, but it also tells a vital story. The story of the cosmic zoom is about the past seventy years of trans-scalar encounter, stretching from the systematizing and disciplining of scientific knowledge to the sublime encounter of ever-smaller and ever-larger forms of radical alterity. We have encountered new scales even as we have solidified our thinking about scale itself. We have encountered the earth as a pale blue dot and discovered that fundamental particles also behave as waves—and do not obey the "standard" laws of physics. We have explored deep space and produced silicon-based ultra-miniaturized gates that have enabled a computational revolution, in turn enabling us to study and characterize a global climate on the edge of a precipitous tipping point. We have experienced the emergence of social media and its datafied and surveilled digital environments. These twentieth- and twenty-first-century trans-scalar encounters constitute an opening up of the milieu of the human so disorienting and awe-inspiring that it might be considered a tear in the space-time continuum, now understood as a scalar spectrum. The cosmic zoom, as a heterogeneous set of medial compositions and as a conditioning of the scalar potentials of the cosmos, has served as a response to this unprecedented situation. It is not, however, entirely new as a conceptual or narrative framework.

In Cicero's "The Dream of Scipio" (*Somnium Scipionis*), the titular character is visited in a dream by his famous grandfather (by adoption), Scipio

Africanus, who takes him on a tour of the cosmos. He first finds himself floating far above the city of Carthage, then ascends higher, until the earth has shrunk to a small globe. His grandfather shows him nine successive spheres, each enclosing the last, which together make up the sum of the universe. Scipio is amazed: "In size the celestial bodies far surpassed the earth. Indeed, the latter was so insignificant by comparison that I was disgusted with our empire, which is but a speck on the surface of the globe."[11]

Despite his awakening sense of scale, the younger Scipio finds, as the cosmic tour continues, that he cannot tear his eyes away from his home planet. His grandfather notices and rebukes his monoscalar fixation: "You are still lost, I see, in the contemplation of that comfortable home of man. If the earth appears to you small, as it really is, keep your gaze riveted upon this Heaven, and care not a straw for earthly things."[12] Scipio Africanus's scalar lesson is simple: Carthage, Rome, and even the entire Roman empire are so diminutive when arrayed against the scale of the universe that nothing of significance has or can be achieved there. Still, however, his grandson experiences this scalar spectrum as radial, anchored, centered upon the earth—which even in this vision is located at the center of the universe. These are the basic ingredients of the cosmic zoom.

Cicero's text emphasizes alterity: the celestial spheres are fundamentally different from the earth, up to and including the "colossal revolutions" that produce "this music of the spheres," so overpowering that "no human ear can endure it."[13] The cosmos is fundamentally alien and incomprehensible to human senses and concepts, attuned as the latter are to a single conditioning scale. Exhorted by his guide to radically alter his scalar perspective, to look outward and experience difference, Scipio cannot abandon his fixation on the point of his departure. To remain fixed upon Earth is, as Scipio Africanus makes clear, to remain fixated upon the human, upon one's own subjectivity, however thickly contextualized spatially and temporally. Scipio's fate is to obtain a view from the cosmos, a view from everywhere, but to remain unchanged, to remain human, all too human. This didactic fable presents us with the roughest diagram of the cosmic zoom. The potential for difference, the trans-scalar encounter, and the reflexive mediation of scale are ultimately collapsed by human subjectivity that seems immune to alterity.

On its surface, the cosmic zoom is simple: it depicts a movement from the smallest known scale of potential experience to the largest (the universe as a whole). The examples analyzed in depth in this book begin with Kees Boeke's book *Cosmic View*, from 1957, and continue through the ground-breaking and extremely influential work of Ray and Charles Eames to cur-

rent cinematic and database-driven media. The cosmic zoom is so ubiquitous in media from the second half of the twentieth century to the present
that it forms something of a master scalar trope. Cosmic-zoom media have,
in large part, *taught us how to think about scale*. Incidentally, they have also
taught us how to think about media, and even thinking itself. The cosmic
zoom is a sandbox for scalar thinking, as will become clear when we view
its constitutive instantiations through a media-archaeological lens. The
cosmic zoom is a reflexive form, with mediation and scale as its entwined
subjects. As I explore throughout this book, the cosmic zoom is more than
a visual trope or narrative technique: it is a scalar ideology, a framework
for ordering the world in relation to the human.

Throughout *The Cosmic Zoom* I treat individual instantiations of the
cosmic zoom as both discursive and material objects. In most cases I explore the processes by which these zoom were constructed as jointly material and conceptual projects. This deconstruction demonstrates both how
scales are stabilized in human knowledge production and how scale dis-
rupts our knowledge practices. But which is true? Is scale a physical property independent of subjective experience, or is it entirely arbitrary, a set
of conventions constructed by discursive practices? In my view, *both* of
these propositions are correct in all but their logical exclusion of each other.
Scale marks both ontological difference that is independent of experience
and arbitrary domains generated by experiential accounts. I refer to this
as the *scalar paradox*, and it will come up again and again in the pages of
this book.

Rather than collapse the scalar paradox, I believe that scale theory demands we hold it open, in productive tension. It is vitally important to
understand scale as a primary ontological determinant of form and function, especially in the face of persistent campaigns in nearly every discipline
toward scalar collapse, or the elision of difference between two or more
scales when they are placed in the same medial frame.[14] Scalar collapse is
the result of epistemological and medial practices that unwittingly or deliberately normalize one scale to the dynamics, features, and cultural status
of another. Collapsing one scale into another is a profitable and productive
enterprise in many fields, and is at this point demanded by global capital
as one of its primary engines of extraction and circulation. In the realm of
thought, scalar collapse takes the form of a naivety or ignorance of scalar
mediation, that is, of the ways in which scales are defined and stabilized out
of manifold material existence, on one hand, and the ways in which matter
differentiates itself into functionally unique entities at different scales, on
the other.

While certain cosmic zooms have been analyzed in passing by many scholars, particularly the most famous and influential instantiation, the 1977 Eames film *Powers of Ten*, the cosmic zoom has never been properly studied as a transmedia project, and, surprisingly, no scholar seems ever to have publicly asked the question in every child's head after a first viewing: "How did they make that?" That the cosmic zoom has never been subjected to an analysis of its own construction, but only analyses of its reception and post facto critiques of its apparent ideology, is a symptom of the biases that hobble past attempts at scale theory in the social sciences and humanities. In this book, I employ a media-archaeological approach to the cosmic zoom in an attempt to uncover the methods, assumptions, and behind-the-scenes struggles that attended the construction of these iconic media works. The purpose is not simply historical curiosity or trivia but rather a far deeper engagement with the scalar paradox itself. As Siegfried Zielinski argues with respect to media archaeology, "The goal is to uncover dynamic moments in the media-archaeological record that abound and revel in heterogeneity and, in this way, to enter into a relationship of tension with various present-day moments, relativize them, and render them more decisive."[15] Every cosmic-zoom project is a battleground of conflicting knowledge practices, the strategic deployment of medial technologies, and an engagement with both sides of the scalar paradox. Excavating the ways that cosmic zooms have been made will therefore afford us the richest possible engagement with the fundamental dynamics of scale.

Before we embark on this fantastic voyage, however, we have to trace the multiple meanings of "scale."

DEFINING SCALE: FOUR DISCIPLINARY MODELS

When we talk about scale, we rely upon long discursive traditions and their attendant assumptions, which in most cases remain tacit. These keep us from fully recognizing Wonderland. Every discipline, academic or lay, has its own understanding of scale. Is scale a core feature of the universe or a way for thought to organize and *represent* the universe? This is the first demarcation line of disciplinary territoriality. Again, in this book I take the view that scale is both: the universe is scaled and scaling in a fundamental way that is independent of human interpretation, but is at the same time stabilized into discrete scales through human knowledge practices. This leads to the scalar paradox of knowledge production: disciplines divide the world into scales, carving up space and time into discrete, simplified compartments that can be studied and manipulated, yet the knowledge they produce cannot help but shatter these scalar-disciplinary boxes. Knowledge produces discrete scales, but knowledge production itself relies upon the dynamics of scale to resolve features of the world. In 1898 Marie Curie discovers radium, an atomic element that appears almost magical in its scalar alterity; sixty-three years later radioactive elements are solidly in the realm of science when technicians at the US Army's SL-1 nuclear reactor, despite following strict protocols, accidentally trigger the first deadly nuclear meltdown. Every circumscription of scale is a foray into a world constantly creating and recreating itself through scalar difference. Every engagement with this radical alterity is an irruption into thought, a reordering of our milieus. To remain disciplined, human knowledge producers isolate and contain those irruptions, relegating them to their proper scales, and continue their work. The meltdowns, however, continue: scalar alterity can never remain fully contained.

Put another way, every discipline has its own way of taming scale. The result is that when we communicate *across* disciplines about scale, our dialogue is confused, piecemeal, and contradictory. When we deploy the concept of *scale*, are we referring to units of measurement, operations of shrinking and enlargement, relative ratios between representational surfaces, absolute size domains, or relationships of force? Do we really know what others, in their own situated knowledge practices, mean by the term? Or for that matter, what *we* mean? Each discipline has its tacit definitions, commitments, tools, and sacred scalar truths. The condition signaled by the concept of the Anthropocene is one in which we are confronted with the dire implications of scale effects at the historical moment of minimum discursive overlap between scale-mediating disciplines. The realization that we should be awakening to is that we lack a critical and shared vocabulary of scale.

This book sets out to rectify this problem. Rather than assuming an uncritical and disciplinary model of scale, rather than exploring scalar dynamics as an important but vague set of implications for some other object of study with which we are all more comfortable, these pages build a tentative transdisciplinary theory and vocabulary of scale itself. The broad scope of this introductory chapter is a response to the challenge implicit in this project: if we are not to rely upon received notions of scale, if we are truly to sound its depths, we must dispense with our preconceptions and problematize the concept itself.

Scale, like sex or identity, functions as an interfacial concept, acting as both an invitation to open-ended multiplicity of encounter and a ready-made, "obvious," even trivial, tool available in our everyday negotiations with the world. This trivialization, this refinement and reduction of the tool of scale, is accomplished by disciplining knowledge production to obey precise boundaries and fit into carefully constructed categories. To problematize scale, to recover its multiplicitous meanings and potentials, we will need to defamiliarize it, to dig under and beyond its trivial definition and ask how it *works*. This requires an explicit engagement with the disciplinary structures that have cordoned off particular aspects of the concept, honing them for ready use in their own domains of knowledge production.[16]

In the remainder of this section I briefly trace four conceptualizations of scale that I feel have the greatest currency in our culture at large, as well as in academic discourse. Most deployments of the scale concept are either straight borrowings of one of these formations or a hybrid of two or more of them. The first is *scale as relational ratio*, derived from cartography. The second is *scale as absolute size domain*, derived from physics. The third is *scale as compositional structure of parts to whole*. This is the dominant understanding of scale in both engineering and biology. Finally, in mathematics scale is generally conceived of as a *homologous scaling operation* by which a figure or pattern is altered in magnitude while holding its internal relationships invariant. Let us explore these often contradictory conceptions of scale one at a time, noting both their critical affordances and their limitations for use outside of their progenitor disciplines.

Scale as relational ratio. Mapping the spatial extension of one's environment is no doubt one of the most ancient human deployments of scale. In its two-dimensional, spatialized, disciplined form—cartography—scale functions as a guarantor of the relation between the map and its territory of interest. Contemporary cartographers refer to scale as the "denominator of representative fraction."[17] Let us parse out this terse definition. Scale is a ratio between distance on the map's plane and distance on the plane of the object of interest (the object being a physical area that has been surveyed in some fashion). The object is "full scale," so its scale is absolute, in whatever unit it is being measured—a numerator of 1. The map, generally smaller than the object, is only fractionally as large. If a map's scale is 4:1, the denominator is 4 and the map is one-quarter the size of its described object. In this case, the given distance between any two points on the map is four times smaller than the distance between the two points of the physical landscape represented by those points. As the above definition indicates, the map's function is assumed by cartographers to be representational; that is, the distances and features signified on the map's surface "represent" corresponding features on the surface of the landscape. The map's scale, along with "contour interval" for topographic maps, determines (fixes) this representational relationship.[18]

Both scale and contour interval are measures of potential detail. Scale, as the ratio between two surfaces, determines how much space can be mapped onto the much smaller cartographic surface. The larger the ratio, the more detail is compressed into a given area of the map; in other words, the more area the map is said to represent. Cartographic scale thus determines both the resolution of the map (the fineness of detail that it can represent) and, in relation to the represented area, the size of the map itself. Contour interval, by contrast, is not expressed as a fraction but as a single value (the difference in elevation between adjacent contour lines). As such, contour interval directly expresses resolution, while scale corresponds to both resolution and map size. Because maps are fundamentally limited in the amount of detail they can reproduce, *scale expresses this tradeoff between size and resolving power.* "There is no map that will fulfill every need."[19] Cartographers, like ecologists, have to choose the "best scale" for their object of interest. While map users often forget about these tradeoffs, except when we can't find what we want,[20] we have nevertheless inherited something of the representational and fractional framework of cartography when we think of scale. In chapter 2, I argue that cartographic scale can be reconceived in nonrepresentational terms as a direct negotiation of ecological detail.

Scale as absolute size domain. Physicists tend to approach scale somewhat differently. In their discipline, scale is usually considered neither representational nor planar (projected onto a flat surface). Rather, it signifies a defined size domain. Planets occupy a certain scale, as do bacteria, electrons, and humans. As in cartography, physicists thus see scales as territories of a sort, only these are more often virtual, generic territories rather than singular ones. Each scale is a conventionally derived slice of reality. Such delineations make what I call a *resolving cut* in order to isolate features of the physical universe that can then be described empirically or theoretically. The concept borrows from Karen Barad's notion of the "agential cut" that, in quantum mechanics, differentiates one region of matter from another, making experience and knowledge possible.[21] To build on Barad's concept, a resolving cut, as a theoretical and as a practical matter, stabilizes a portion of the scalar spectrum, isolating a particular scale. It does so through a medial apparatus that determines what features become legible or readable for the assemblage making the cut. Like Barad's agential cut, a resolving cut is a differentiation of time and space from within; in this case, it constructs a relationship between two distinct regions of the scalar spectrum. Any resolving cut is in one sense arbitrary, but in resolving the difference between the surface of observation and the surface upon which trans-scalar details appear, it enables fundamental ontological difference to emerge. The production of scales, for humans in particular, is thus inseparable from the differential functioning of disciplines, fields, and subfields. These knowledge domains come into focus through the resolving of specific material scales, whatever they may be. I explore these dynamics in further detail in chapters 4 and 5.

Scales cannot, of course, be usefully defined by specific Cartesian coordinates, as they must function *generically*. They are size domains, not determinate spaces. This is why the difference between size and scale—two concepts that are commonly conflated—is of great importance. The nanoscale can be as small as the head of a pin or as large as a galaxy; its spatial extension is arbitrary. What marks it as the nanoscale is the typical or characteristic size domain of its entities and dynamics. The nanoscale is the domain in which features measured in nanometers can be resolved as individual entities. Size is absolute and subject to direct measurement by the physicist. Scale, on the other hand, is relative: it requires that a relationship be stabilized between at least two entities. Scale already, then, smuggles in this process of stabilization itself. Scale is reflexive, size is not.

The physical notion of scale thus implicitly incorporates a notion of field of view, a perspectival phenomenon: what sort of entities are resolvable in a field of view only a few tens or hundreds of nanometers wide? In this sense, scale, unlike size, always exceeds the disciplinary apparatus that frames it; we stabilize scales, but we never know what we will find as a result of the encounters we thereby initiate. Scales are speculative: they define a field of view in which entities then become resolvable, often through technological mediation, as when the Dutch inventors of the microscope peered through it and began to describe the entities it made visible, giving birth to a new discipline and a new, stabilized scale. Of course, this scale existed in more fluid form prior to its disciplining in science, as a speculative milieu of germ cells, homunculi, and so on. As we shall see in chapters 2, 5, and 6, scales can be virtual ecologies even subsequent to their disciplinarization, when speculative media conjoin objects that occupy different spaces but similar scales.

In its production of new scales, physics proffers not only new entities, but also new insights into differentiation itself. Difference exists not only between an entity and its representation, or laterally between entities, but also along a scalar axis. To take a prosaic example, nonphysicists commonly assume that gold is straightforwardly yellow: no matter how tiny or how large the pile of gold, it will gleam with that hue. But at the nanoscale, gold can be orange, purple, red, or green.[22] Physics teaches us that yellow is a *macro* quality of gold, not a scale-invariant quality. The very same coordinates in space, resolved by an atomic force microscope, an electron microscope, an optical microscope, the naked human eye, and, at a distance of some light years, the Hubble Space Telescope, would be revealed to contain different entities, each unique in its characteristics, as well as the dynamics that form between them, the forces that affect them, and their capacities for structuration, deformation, complexification, and differentiation. These are scalar strata that occupy the same space but not the same scale, and can only be revealed through one or more forms of mediation. As I argue in chapter 5, such engagements necessitate a new understanding of the relation between scale and difference.

Despite the fecund results of physics' scalar differentiations, however, a countervailing tendency within the discipline has sought to contain such eruptions of difference through the positing of a homogeneous and holistic universal model governed by a single logic. The most notorious example of such an end run around scale within the discipline of physics is no doubt Isaac Newton's universal laws of motion and gravitation, commonly referred to collectively as "the clockwork universe." This conception of a deterministic universe requires and thus suggests—without evidence—that scalar difference is essentially illusory, that all entities at all scales behave in exactly the same way. This is the meaning of the likely apocryphal apple story: Newton sees an apple fall and "realizes" that apples and planets are fundamentally alike. While Newton's metaphysics (and thereby his physics) were determined by his particular theological predilections, as I briefly consider in chapter 5, many physicists have similarly labored to promote a model of the physical universe that would belie its apparent difference—as revealed through empirical observation—and repackage it as a kind of scalar layer cake in which each scale contains different entities, but all are unified and homogenized within the cake form.[23] This quest has consumed many physicists, from Albert Einstein, who defiantly declared in the face of quantum indeterminacy that "*He* [God] is not playing at dice," to David Bohm, who acknowledges the discontinuities of the quantum but posits a mysterious and metaphysical "implicate order" underlying them, to Stephen Hawking, whose mythical "complete unified theory" would homogenize the dynamics of all scales into "a complete description of the universe we live in."[24] These individual physicists, as well as many others, have embarked on quests to unveil scalar difference as an illusion or matter of perspective rather than an ontological fact. Einstein's concept of relativity, for instance, seeks to remove time as an ontological determinant, collapsing it into space (technically "space-time"). As we shall explore in chapter 3, when the Eames Office set out to present a unified medial instantiation of a scale-free universe, they invoked Einstein to legitimate the frictionless space of perspectival mastery they were in the process of constructing. The projects of Newton, Einstein, Bohm, and Hawking, however influential, have all conspicuously failed to account for the totality of the scalar spectrum, each demonstrably breaking down at certain scales. The universe seems to be fundamentally scale-discontinuous.

Scale as compositional structure. In the biological sciences, as well as in engineering, the central scalar problem is that of function. How do organisms, bridges, and skyscrapers function differently as their size increases or decreases? Here, the effects of scale on an organism's or engineered structure's ability to sustain its form and successfully interact with its environment is one of the central objects of knowledge production. In biology, this problem is called "allometry," and describes the limitations imposed by scale on relations between an organism's whole (body) and its parts (organs). While these relations can be described by ratios between organs, skeletal structures, and so on, the ratios are not representational (as in cartography) and change as the absolute size of an organism changes. The central insight here is that organisms do not linearly scale: any change in size requires a redistribution of organs and their functions—in other words, a redesign of the organism. As biologist D'Arcy Thompson expresses it in his classic tome *On Growth and Form*, "There is an essential difference in kind between the phenomena of form in the larger and the smaller organisms."[25] This is due to the relative efficacy of forces at different scales (physics) as well as problematics of surface area versus volume, which affect how oxygen and nutrients are diffused through tissues. As biologist John Tyler Bonner notes, "Size is volume, yet life's activities require the appropriate surface to go with the volume and the result will be different shapes for different sizes."[26] Structures that work to distribute nutrients, gases, and waste within a tiny insect (such as direct tubules) will not work for larger mammals, which require increasingly complex circulatory systems utilizing blood as their medium and entail great increases in relative surface area and complexity as they scale up.

For biology, then, scale describes the constitutive relationships between size, parts, and whole. These relationships, moreover, reconstitute themselves at multiple scales: the cell integrates its parts in a fashion particular to its scale, and functions in unison with many other cells to compose an organism whose structure is keyed to *its* particular scale, and so on. Ultimately, the demands of scale on structure produce deep structural discontinuities within a physically (even organically) contiguous spectrum. As Thompson notes:

> In the end we begin to see that there are discontinuities in the scale, defining phases in which different forces predominate and different conditions prevail. . . . Man is ruled by gravitation, and rests on mother earth. A water-beetle finds the surface of a pool a matter of life and death, a perilous entanglement or an indispensable support. In a third world, where the bacillus lives, gravitation is forgotten, and the viscosity of the liquid . . . the molecular shocks of the Brownian movement, doubtless also the electric charges of the ionised medium, make up the physical environment and have their potent and immediate influence on the organism. The predominant factors are no longer those of our scale; we have come to the edge of a world of which we have no experience, and where all our preconceptions must be recast.[27]

Inevitably, then, a full consideration of scale in biology leads us to the problematics of topology and environment, or ecology. Here it is clear that structural scale in the world of the biologist or engineer is inseparable from the size domains of physics, even if it reveals different details and implications of scale through its divergent disciplinary problematics.

Scale as homologous transformation. Mathematics, on the other hand, when severed from empiricism, has tended to employ scale in a frictionless environment. Size domains do not apply to pure mathematics, nor do the constraints of changing parts-to-whole relations. Geometry's parts-to-whole relationships are scale-free: as a geometric shape is scaled up or down, its internal angles and ratios remain invariant. The universe of geometry, then, is wholly unlike that explored in cartography, physics, biology, or engineering. Here, *lack* of scale is its most salient feature, the master principle that ensures its consistency and thus its coherence. Mathematics, we might say, is without ecology—even if the reverse does not hold.

Frictionless scaling is not limited to geometry: all mathematical functions are, at their heart, abstract machines that exist in a frictionless environment of continuous correspondences between variables. This is to say that, in its essence, the function describes a continuous abstract space that generates, for any input, a corresponding output, the totality of which describes or at least implies an unbroken, graphable correspondence between homologous points, whether as a linear line, a sinusoid, an asymptote, or some other form. In its essence, mathematics describes continuities, despite its use of discrete digits. Of course, because the actual universe is chock-full of discontinuities, applied mathematics must attempt to model this, as in quantum mechanics and nonlinear dynamics. The point is not that mathematics is incapable of describing discontinuous relationships but rather that it is at heart, before applying the constraints introduced by empirical description, the medium of a virtual, frictionless world of infinite capacity. As Gilbert Simondon argues, "Theoretical thought that makes use of numbers is essentially contemplative and of religious origin. It does not seek to count or measure beings, but to estimate what they are in their essence in relation to the totality of the world."[28] Mathematics transposes the physical into the realm of the metaphysical, where abstract forms rather than actual entities are compared without friction.

It is precisely this boundless flexibility, this capacity to surpass, subsume, and describe any bounded system (the real) that has thrilled mathematicians since antiquity. In the third century BCE, Archimedes wrote a treatise, "The Sand Reckoner," nominally addressed to the Syrian king Gelon, in which he gleefully calculates the upper bound for the number of grains of sand in the universe. Sand itself had been synonymous with the "uncountable," an incalculable quantity. In seeking to overcome this implicit trumping of number by matter, Archimedes first measures the number of grains of sand that fit within the diameter of a poppy seed, and from there begins to multiply the result by larger and larger units of measure until he reaches the size of the entire universe, as estimated by Aristarchus—where "universe" is understood to be the largest sphere of the cosmos that holds the invariant stars and contains the other heavenly bodies. To accomplish this, Archimedes invents a new system of numerical notation. Whereas the largest named number in the Greek system was the "myriad" (ten thousand), and thus the largest denotable number was a "myriad myriad" (one hundred million), Archimedes uses the myriad as a base, developing an exponential system, which allows him to add ordinal "places" rather than multiplying increasingly unwieldy sums. Using the volumetric upper bounds of such bodies as the earth, the sun, and the universe, as calculated by mathematicians of his day, he eventually concludes that the total number of grains of sand that *could* fit in the universe must be less than "10,000,000 units of *eighth order* of numbers" (10^{63} grains). Archimedes acknowledges that "these things . . . will appear incredible to the great majority of people who have not studied mathematics," but of course, for those who have, his argument will be recognized as a powerful proof.[29]

The key to this text, however, is not that it enables Archimedes to calculate the actual number of grains of sand in the universe—he is wholly uninterested in this empirical question—but rather that it allows him to establish the greatest *possible* number. Should physicists alter their model, making the universe larger, Archimedes's notation system enables the mathematician to simply increase the ordinal number (exponent), adding one to effect an order of magnitude jump in size. Let's see sand keep up with that scalar technique! His proof, of course, concerns not sand but mathematics itself: number, it avers, is capable of trumping any actually existing quantity of anything. This reversal of priority between number and matter is the decisive ideological turning point for mathematics. Mathematics, using Archimedes's exponential system, can produce infinitely scaling forms: there is no limit to the magnitude it can signify or the system it can model. In one fell swoop, Archimedes has produced a system to derive arbitrarily large numbers and make them practically calculable, and has thus produced a way of mathematically "zooming" from one arbitrarily sized entity to any other arbitrarily sized entity. He thus lays into place the foundational mathematical, philosophical, and ideological tools for the twentieth century's cosmic zoom.

Archimedes's victory is a dangerous one. Freeing mathematics from empirical constraints is key to its functioning in theoretical domains, and certainly the source of much of its mystique and prestige, up to and including the contemporary era. However, we must take care not to backport its scale-free models to the world of difference, interdependence, and interaction that we actually inhabit. The history of mathematics and science is replete with such attempts. Already before Archimedes, Pythagoras had suggested that the cosmos was structured according to perfect ratios of whole numbers. Empirical observation to the contrary would not dissuade him. He even went so far as to suggest that, because ten was the perfect number, there must be exactly ten heavenly bodies in the universe. Only nine could be observed, so he invented a tenth: an anti-earth.[30] This is not to impugn theoretical cosmology or physics, but merely to suggest that it is all too easy to make oneself believe that just because something is possible mathematically it is possible physically. Such slides become even easier when we, like Pythagoras, subscribe to a form of aesthetics derived from mathematical proportion; this is the primary conduit by which mathematical forms come to seem more real than empirically observable reality, as Plato's theories of geometrical atomism (elaborated in his *Timaeus*) and that of the Forms (elaborated in book 7 of the *Republic*) attest.[31] This mathematically derived aesthetics is precisely an aesthetics of *freescaling*. Its most contemporary form is perhaps that of fractal geometry, which dispenses entirely with Pythagoras's whole number ratios, and recuperates a sense of the infinite and uncountable in natural forms (the contours of any coastline, to take a celebrated example), yet still seeks to reproduce a freescaling aesthetics. Benoit Mandelbrot informs us that "many of the irregular and fragmented patterns around us," which he names fractals, "tend to be *scaling,* implying that the degree of their irregularity and/or fragmentation is identical at all scales."[32]

The very limited set of truly fractal patterns in the world is sometimes taken, in popular culture, to be a sort of general model of scale-invariance, incorrectly regarded as applicable to nearly any phenomenon. It is important to keep in mind that only certain phenomena exhibit fractal properties and that even for those, *not all of their properties* are scale-invariant, just their degree of boundary irregularity. Many structures *appear* similar at different scales, but we must temper our reflexive declaration of "self-similarity" by specifying a set of relevant dynamics: which relationships are actually self-similar across scales? It is all too easy to conflate visual similarity with actual mathematical homology, which requires a set of points in identical relationships at different scales or precise patterns that repeat at different scales. Computer-generated fractals, which scale infinitely, here seem to act as a kind of hypnotic trigger, giving us the same sense of infinite scaling imparted by Euclid's forms, which serve to bolster human fantasies of transcendence.[33] This impulse places the human in a privileged position of access with regard to the universe, which implies both a mastery and a self-centrality as a species—as privileged, rational, mathematical creatures. The first step toward the illusion of transcendence—another magic trick—is always to conquer or occlude scalar difference.

TRANSDISCIPLINARY KNOWLEDGE PRODUCTION IN THE ANTHROPOCENE

If we wish, as a culture and as a species, to increase our scale literacy, it will require a self-reflexive engagement with all four of the disciplinary formulations of scale explored in the previous section. Biology and engineering provision us with scale vis-à-vis the object, reminding us that scalar difference is irreducible: differently sized objects function differently. That is, the relation between parts and whole necessarily change along with an object's size. Godzilla and fractals aside, scale is not a secondary quality that can be applied to an object without changing its essence; rather, scale marks a spectrum of discontinuous constraints on the organizational possibilities of assemblages. Along with the size domains of physics, these restraints remind us that resolving cuts give us domains that have different rules. We have not imposed these rules upon them; they impose them upon us whenever we encounter them. This is, in some sense, the first step away from an anthropocentric orientation toward scale. The size domains of physics take us further, beyond the object as such to scale as relationality, as a set of dynamics and potentials that are once again singular to each scale, and thus to each resolving cut we make. The scaling operations described by mathematics are equally important to confront, as they have been the most weaponized against our ecological milieus, principally in the form of petro-capitalism, when they could instead be applied to the virtual dimensions of scale: scalar transformations that potentiate new milieu dynamics. Cartographic scale provides the necessary emphasis upon the process by which resolving cuts are made in the first place. That is, cartographic ratio confronts us with our own medial systems and, when divested of its representational framework (as we shall see in chapter 2), also forces us to confront the relationship between resolution and perceivable detail. Cartography thereby brings us face to face with the fundamental process of negotiation by which scales are stabilized for observers.

An ecological view requires a partial integration of these disciplinarily divided forms of scale. Resolving cuts, the stabilization of scale domains, ontological scalar difference, and scale constraints on the composition of assemblages are all related and must be thought together. In this sense, we have much to gain by unthinking some of the boundaries and purifications that have honed these concepts for particular disciplinary purposes. We must see scale in its full light, as ontological difference, construct of knowledge, and speculative ecology—all at once.

Outside the disciplines of mathematics and the natural sciences, knowledge practitioners have picked up these various notions of scale and integrated them into their own strange brews according to their needs and predilections. In computer science, the increasing scale of data to be stored, manipulated, and searched has led to formulating scale as processing efficiency (algorithms) rather than raw processing power (hardware clock cycles). The corporation that most successfully formulated and tackled the problem of data scale on the early Web, Google, did so by developing the most efficient search algorithms and distributed computing infrastructure. The problem, however, remains a perennial one for the discipline as the scales of networked data proliferate and grow. As we will explore in chapter 6, the problem of big data has become a problem of subjectivity itself.

The social sciences, no less than computer science, have increasingly attempted to integrate scalar concepts into their practices throughout the twentieth century. Psychology, structural anthropology, linguistics, and sociology have all sought to discover patterns that hold across scales, and more importantly, explain the relations between phenomena observed in smaller and larger domains. In the field of historiography, the Annales School—and especially its superstar practitioner, Fernand Braudel—

developed a form of large-scale historiography (*longue durée*) that analyzed history as the product of environmental and socioeconomic processes that play out in patterns over millennia rather than in the short-term actions of individuals.[34] Manuel De Landa rekindled this torch in modified form, pushing further into nonhuman territory with his monumental *A Thousand Years of Nonlinear History*.[35] Human geography similarly attempts to expand the scale at which we resolve humans' interactions with their environment, analyzing them in spatial terms. After the late 1960s, however, intellectuals increasingly focused on the processes by which conceptions of space were *produced* by humans, and how these conceptions ordered actual spatial distributions. Michel Foucault's genealogies of power traced the connection between knowledge production and spatial distribution through exemplary studies in singular and circumscribed spaces of power that span scales as diagrammatic compositions. These diagrams of power order entire populaces via the topological and categorical structures they stabilize: the hospital in *The Birth of the Clinic*, the asylum in *Madness and Civilization*, and the prison in *Discipline and Punish*. Around the same time, Henri LeFebvre published his groundbreaking *The Production of Space*, which analyzed the history of spatial ordering as a dialectic between centers and peripheries leading to ever more abstract spaces determined by the exigencies of capitalism.[36] These works, which drove the "spatial turn" in the social sciences and humanities, employed the concept of scale not only as increasingly large size domains, but also as a form of large-scale or collective cartography that produced and reproduced the conditions of individual experience and identity. Society was not only a series of movements larger than its constituent actors but also the outcome of an emergent form of spatial production that functioned on multiple scales simultaneously. These discursive frames helped blur the lines between individual actors, larger-scale social dynamics, and infrastructural and environmental determinates, thus opening a conceptual space for interdisciplinary collaboration between the social sciences and humanities.

The traditional humanities have had a more difficult time shifting the resolution of their analyses to larger scales, though historiographical movements in the digital humanities have more recently advanced programs of "distant reading" or "macroanalysis" that do precisely this, shifting the analysis of cultural history toward the quantitative and large-scale, several decades after the initiation of the spatial turn.[37]

Another relatively recent approach paving the way for a (post)humanities theorization of scale relying on qualitative rather than quantitative analysis falls under the rather wide umbrella of the "material turn." Like the spatial turn, this implicit repudiation of the linguistic turn that enamored the humanities in the 1970s through the 1990s purports to take matter and materiality seriously. We may roughly divide this movement into the object-oriented ontologists and the new materialists. The former have elaborated what Levi Bryant calls a "flat ontology" with only one class of matter: objects.[38] Timothy Morton's influential book *Hyperobjects* approaches the scalar concerns of the present as a "being-quake" heralded by the appearance of "hyperobjects" that massively outscale the human.[39] While I am sympathetic to this stringent appeal to objectness, *Cosmic Zoom* owes a much greater debt to the new materialists, who place less emphasis on objects and more on processes of becoming. Growing out of the materialist vitalism of Friedrich Nietzsche, Henri Bergson, and Gilles Deleuze, new materialists such as Jane Bennett and Karen Barad theorize matter itself as possessing active, compositional qualities. As we shall see in subsequent chapters, the lively (Bennett) and agential (Barad) qualities of matter help us make sense of the co-constitutive nature of observer and object in encounters across scalar difference, without recourse to human subjectivity.[40]

While the humanities may have come to the "scale table" fashionably (or just embarrassingly) late, their greatest disciplinary asset is perhaps their expertise in the self-reflexive analysis of knowledge production, including the intertwined processes of ideological reproduction, cultural narrative, and identity formation. While other disciplines have contributed enormously to knowledge of and about difference at disparate scales, and human processes as they are imbricated in large-scale processes, the humanities possess three unique potentials for scale theory. First, they are capable of problematizing scale itself as it is deployed in our culture—that is, rendering the specific toolset of scale visible, foregrounding it as a problem in itself. Second, their discursive analytics, accustomed to analyzing the ways knowledge is produced in many disciplines, could potentially integrate different discourses and disciplinary tools into a transdisciplinary form of inquiry and collaboration—but only if they can overcome their own disciplinary shortcomings. Finally, the humanities possess an accumulated expertise in the virtual dimensions of culture, and thus the capacity to shift the focus from an empirical study of entities and processes at different scales to the production of new forms of scalar thinking and engagement. Scale, in the hands of the humanities, could become more than a domain of analysis or critique; it could become a new horizon of thought, a collective project commensurate with the novel challenges of the Anthropocene.

We use scale to order our knowledge of the world even as the world uses scale to embarrass our orderings. These are the two sides of scalar mediation. Any proper study of scale must confront this aporia: we make scales; scale unmakes us. Despite attempts by humanist, capitalist, and industrialist apologists to contain the scalar problems of climate change, mass extinction, ocean acidification, and looming resource shortages within logics of market-based solutions, national regulation, and "clean" technology development, it has become increasingly clear that our regimes of knowledge production are deeply imbricated in the production and strategic occlusion of scalar relationships. The problem isn't that we don't talk about other scales (we do this constantly) or that we don't have medial access to them (we have unprecedented access in many media). The problem is that our access is mediated in such a way as to delimit our tactical responses, to frame and contain our conceptual engagement. The problem isn't that we lack effective scalar media, but that our scalar media are *too* effective—at predetermining the shape of our encounters. Any cultural scalar analysis must therefore begin by charting how scales are *stabilized* (speculatively, rationally, disciplinarily) both initially and over time in response to perturbations and challenges. Thus scalar problems like climate change and mass surveillance cannot be solved as long as they are framed as problems only within and by the forms of knowledge that have given rise to these dynamics. Rather, we must reconceptualize these problems from new, self-reflexive positions that take scale not as a given set of preexisting relationships but as a dynamic of mediation, and thus as a set of ongoing negotiations.

Access to scalar alterity requires resolving different scales, which in turn demands techniques by which one surface is put into contact with another. Media theory offers us vital tools in this analytic, not the least of which is that of the "affordance," or materially enabled potential contained within any medial form,[41] which allows us to evaluate the parameters of encounter with alterity in regimes of trans-scalar access. Ultimately—and here we return to the fundamental scalar paradox—all scalar mediations that frame access to other scales ultimately lead to encounters that, however circumscribed, push back against those frames. Thus scalar mediation can be viewed from two perspectives or stages (though neither of these stages is temporally prior to the other): first, the framing, through knowledge and technical affordances, of regimes of access to other scales and, second, the continual, primal differentiation and composition of elements into entities

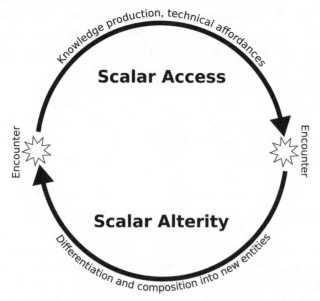

Figure 1 ▸ The circuit of scalar mediation.

at new scales. Together, these mediations form an uneasy, co-constitutive circuit (figure 1).

We experience both sides or vectors in this cycle. The first we experience as our own stabilizing knowledge practices, which order and narrativize our conditions of experience into defined (if implicit) milieus. The second we experience as the impingement or perturbation of other assemblages upon us, which constantly modifies our milieus. This pushback of scalar alterity exceeds our framing narratives and stabilized scalar boundaries, but it is experienced by us only as a result of those regimes of access. Disturbingly, they haunt and harass us as destabilizing, nonhuman agency. Whether as climate change, a viral infection, a declined credit card, quantum entanglement, or bioaccumulated toxic particles, these "strange strangers" jolt us into the realization that others mediate scale just as we do.[42]

Media, as particular conceptual assemblages, provide us with many of the tools needed to theorize this cycle of trans-scalar encounter. Media theory is useful in part due to its historical focus on materiality (the substrata or mediums in which communication is embedded), embodiment (how meaning is generated through interfacial structures and protocols), environment (the navigation and alteration of terrain), and technics (the protocols and affordances of particular technological assemblages). It allows us to understand how thought and matter are connected, and to trace the formation of assemblages that transduce matter and meaning between them. This was implicit in Marshall McLuhan's original formulation of the media concept: "The personal and social consequences of any medium—that is, of any extension of ourselves—result from the new scale that is introduced into our affairs by each extension of ourselves, or by any new technology."[43] Media articulate new scales, which in turn usher in new constraints and potential becomings. As should be clear by now, however, McLuhan's anthropocentric containment of the media concept is too limited. Nonhumans produce media as well, from buckyball molecules to orca pods to the dark matter that pervades the universe. A media theory that answers what I call the "trans-scalar challenge of ecology" must work at all scales without collapsing their difference, and must extend itself not only to nonhuman animals but to nonorganic entities.[44] It must go all the way around the circuit of scalar mediation.

While media theory has plenty to teach us about scale, scale also has a lot to teach us about media. One of the central arguments of this text is that scalar differentiation *is* mediation. That is, we can understand scale as a form of mediation that paradoxically engages fundamental scalar alterity as negotiated surface differentials but also produces certain milieus based upon scalar stabilizations. Media link ecology to subjectivity: a milieu (medium) is "made up" of scalar difference, which stabilizes the subject that narrativizes it. James Gibson's environmental affordances are revealed and activated by what I have been calling *resolving cuts*, or stabilized slices of an environment that produce a scale and thus determine what level of differential detail can be revealed upon its surfaces. This process takes place as a negotiation between a primary and a secondary (medial) surface, and is thus simultaneously technologically and conceptually mediated. This theorization of scalar mediation, developed more fully in the next chapter, reveals, from one angle, scale as a product of milieus and, from another, milieus as products of scale. A milieu is a signifying environment constrained by resolving cuts that nevertheless reveals itself continually to be a *shared* space simultaneously resolved by scalar others.

These considerations push us to expand our notion of media beyond the terms by which it is understood colloquially, as well as its most common academic forms. Media cannot be reduced to communication channels for information transfer or to a certain class of technologies. Neither should media be understood only as an established regime for the dissemination of mass communication (the "mass media" or "news media") or as platforms for the organization of semiotic content. Media are all of these things, but they are much more besides: they constitute a system within which we are already enmeshed, not as consumers, and not even as producers, but as nodes in an environment. This medial environment exceeds the anthropogenic: it is machinically heterogenic, in Félix Guattari's sense of self-generating and -differentiating, and forms operationally closed but constantly evolving assemblages, in the sense captured by Matthew Fuller's concept of "media ecologies."[45] My emphasis throughout this book is, in addition, on the affordances of both human and nonhuman medial systems for interaction with actual (not represented) environments that co-determine the forms of its assemblages.

My focus is not only on the description of medial forms—an empirical practice—but on medial futures: the virtual dimension of media assemblages. This is to consider how media produce new relations that exist outside of themselves and their protocols. Sarah Kember and Joanna Zylinska have figured this conceptual shift from media as objects to media as processes of becoming as a conceptual move from "media" to "mediation." More than simply media in action, "*mediation can also serve as a name for the dynamic essence of media*, which is always that of becoming, of bringing-forth and creation."[46] While I do not insist on a hard and fast distinction between "media" and "mediation" as signifiers, the distinction Kember and Zylinska develop between "representationalist aspirations aimed at closing the gap between the viewer and the screen" and a mode of medial "lifeness" is a key component of medial scale theory. Scalar collapse is, at the medial level, driven by the received (and actively promoted) view that media are fundamentally representational, that they stand in for something else. Because all media mediate scale—that is, they stand in the middle of at least two scales, producing effects across a scalar boundary—a representationalist view of media reduces this mediation to a mere scale model: the television shrinks a recorded image of something else to the size of the screen, a map represents a larger territory, and so on. We are already quite aware of the inadequacy of this conception—we know that the television does far more than passively echo an event elsewhere, and that a map produces or stabilizes a territory rather than passively denoting it. We simply haven't fully applied these insights to the scalar dynamics of mediation. An adequate theory of scalar mediation must therefore be nonrepresentational. That is, it must be able to capture how assemblages come to be different, not merely how one system represents another.

Scale, as a form of mediation, is both spatial and temporal. Human media tend to stabilize spatial scales easier than temporal ones, leading many of us to visualize or conceive of scale as primarily spatial. Yet the cosmic zoom, as a form of media and form of thought, always marshals differential timescales as well as spatial scales. As we shall see, particularly in chapters 3 and 4, timescales and spatial scales are co-constitutive, an ontological fact that has proved convenient to cosmic-zoom designers who have wished to exchange time for space in order to more effectively suture their real-time human viewers to an implied, often universal, observer. A focus on scale as mediation helps to remind us that even in medial systems that explicitly evoke only spatial scaling, temporal scaling is always taking place, even if only "behind the scenes." Chapter 6 considers the linkage of spatial and temporal scale in the context of database-driven media as what I call the "generic scalar event."

The often tragic interlinking of temporal and spatial scale is familiar to anyone concerned with the problematics of the Anthropocene, the era in which we chart not only the spatial scale of damage wrought by our species, but also its temporal reach. As a geological epoch (official or unofficial), the Anthropocene marks a self-reflexive awareness, the moment when the human enters deep time and makes its unflattering mark. Jussi Parikka has suggested that the concept of media is particularly suited to geological thinking and analytics. In his view we ought to extend our concept of mediation to include the forces of the earth at expanded temporal and spatial scales. We must not continue to discount the geological and ecological substances and processes that underlie our contemporary media ecologies. "Nature affords and bears the weight of media culture, from metals and minerals to its waste load." Thus, tracing "the affordances that enable digital media to exist as a materially complex and politically . . . mediated realm of production and process" must become part of our medial narrative.[47] Yet, even as media theory extends the notion of media to geological processes, geology suggests a trajectory for media studies that radically decenters it from the axis of the human. Geological processes produce media before and beyond the human, at radically different temporal and spatial scales. "Geology becomes a way to investigate materiality of the technological media world. It becomes a conceptual trajectory, a creative intervention to the cultural history of the contemporary."[48]

Similarly, John Durham Peters's *The Marvelous Clouds* calls for a renewed engagement of media theory with nature, "the background to all possible meaning." Media, as "ensembles of natural elements and human craft," provide the infrastructural ground for being.[49] Air, earth, fire, and water are natural mediums within which we devise means of surviving and communicating. Each has its affordances and has given rise to different infrastructural formations at the intersection of environment (ground) and techné. Peters urges media scholars to contextualize digital communication technologies in relation to the larger infrastructural assemblages that they exist within, depend upon, and modify. In the same vein, nonhumans possess rich media, extending far into areas that, in the human context, we consider technology: "Technology to humans is nature to many animals."[50] These calls to think technique beside technology, the nonhuman beside the human, and elemental mediums beside the digital, point toward a media studies that ventures far beyond the scales of the human and the humanness of scale. Peters's meditation on elemental media reengages both vibrant matter (in Bennett's sense) and medial alterity, opening an inquiry into grounds and potentials. "In the grandest view," he reminds us, "media studies is a general meditation on conditions."[51] Perhaps not coincidentally, this echoes Ernst Haeckel's founding definition of ecology: "the whole science of the relations of the organism to the environment including, in the broad sense, all the 'conditions of existence.'"[52] We may wish to say that media make up part of ecology just as much as ecology is a form of mediation. Scalar mediation articulates these into the same conceptual circuit and material network.

Alexander Galloway and Eugene Thacker argue that "networks are a matter of scaling, but a scaling for which both the 'nothing' of the network and the 'universe' of the network are impossible to depict. One is never simply inside or outside a network; one is never simply 'at the level of' a network."[53] Digital networks are not networks *of* things, they are networks as processes, as consequential mediations. One of their primary effects is asynchronous scaling, even as networks evade scalar coordinates themselves. The network form, precisely because it enables rapid scaling without itself possessing a determinate scale, becomes the key infrastructure of neoliberal capital, and thus the chief conjurer of what I above referred to as "scalar magic." In this book I engage the network form not only as an object of analysis, in chapters 5 and 6, but as a mode of visualization, in chapters 4 through 6. Here I utilize network graphs to visualize the interrelations between scalar milieus produced by cosmic-zoom films. The network graph serves in this capacity to explore trans-scalar ecology and reveal scalar biases that are occluded by the mesmeric aesthetics and temporal manipulations of the films themselves. The network graph thereby acts as a simultaneous defamiliarization and demystification of time-based cosmic-zoom media.

Beyond the scaling affordances of digital media and their consequent efficacy in contemporary control regimes, I also wish to consider a second axis of digitality that arises as a process, not of computational transformation (scaling), but of scalar stabilization. This axis of digitality marks not the division between continuous contour (the analog) and discrete code (the digital) but rather that between scalar contiguity and discrete resolving cuts. We might call this *scalar digitality* in order to distinguish it from computational digitality, on one hand, and scalar continuity/contiguity, on the other. N. Katherine Hayles's concept of "intermediality" is a useful tool for thinking through the "complex transactions between bodies and texts as well as between different forms of media. Technological functions (making, storing, transmitting) understood as media effects."[54] In this book I consider scale as a primary axis of intermediality. Scale is, in this sense, both an intermedial effect and a primary site of intermediation. While traditional differentiations between analog and digital media rely upon the distinction between discrete and continuous communication channels or mediums, when we consider scalar media *effects*, the medial construction of continuous scalar gradients (in "analog" media) actually works to occlude difference and negate the resolving cuts made by mediation. At the same time, intermedial negotiations generate resolving cuts with profound implications. In both modes or directions of intermediality, the site of translation between the digital and the analog is a key site of scalar mediation: it is here, where one assemblage is encoded or decoded in negotiation with another, that the trans-scalar encounter most often takes place.

While all forms of media bring two or more surfaces into interfacial relation, and the process of mediation that establishes their relationship must always grapple with the differential position of those surfaces along the scalar spectrum, that encountered alterity can be resolved for an observer as either discrete detail or contiguous scalar analogy. I refer to the first case as "scalar digitality" and the second as "scalar continuity," a form of analog scale. As we shall see, what we think of as analog media is capable of mediating scale digitally, and digital media often mediate scale as analog. This study, then, seeks to complicate our common notions of analog and digital media as well as to trouble our received notions of scale.

PROGRESSION OF THE BOOK

The modern form of the cosmic zoom begins with a little book written by radical Dutch headmaster Kees Boeke in 1957. Following the theoretical considerations of the scale concept covered in this introduction, chapter 2 explores the conceptual and material apparatuses of Boeke's remarkable work, *Cosmic View: The Universe in Forty Jumps*, tracing the political, historical, and technological forces that converged in its construction. I will flesh out the concept of scalar mediation by exploring the work's self-reflexive materiality, which it deploys to establish and then stabilize a series of relationships between surfaces—for instance, a page of the book and the courtyard of Boeke's school. This discussion leads to a tentative definition of scale as a processual (medial) negotiation of difference between surfaces. I discuss the mediating relations between surface and milieu, on one hand, and mediation and resolution, on the other. Along with a robust account of scalar mediation as a material practice, one of my central concerns in chapter 2 is an exploration of the concept of *resolution* as it relates to scale and media. Reading *Cosmic View* as a "drama of resolution," I argue that the discursive framework of scale and the medial practice of resolution render scalar mediation a simultaneously ecological, narrative, and onto-genetic process. Here I characterize the trans-scalar encounter as a narra-tive excursion into resolved ecological difference between discontinuous points on the scalar spectrum, necessitating an accretive scalar memory. Each trans-scalar encounter, however, is also an encounter with mediation itself—a theme that develops throughout this book, with differential impli-cations in analog and digital media.

Chapters 3 and 4 go behind the scenes to excavate the genealogies that ultimately produced the most influential of cosmic zooms, *Powers of Ten*, a short film by Ray and Charles Eames. Chapter 3 treats a selection of the Eameses' earlier "toy" and "computer" films as explorations of scalar dy-namics, material encounter, and information theory. These projects culmi-nate in 1968's *Rough Sketch*, their first full-scalar-spectrum cosmic zoom.[55] As with the earlier book *Cosmic View* and the later film *Powers of Ten*, I treat this project not as a post facto media object but as a process of mediation with two phases: its fraught construction and its later intermedial compo-sitions. From aesthetics to the history of science and optical technology to questions of interface and temporality, I examine many of its influences, as well as the intense disagreements and agonizing decisions that marked the construction of the film by its core development team. As in other chapters, I employ archival research that challenges received views about Boeke's and the Eameses' projects. Chapter 3 focuses on the aesthetics of the zoom as a medial process and trope, as well as the attendant politics of animating trans-scalar encounter. Building on the scalar stabilizations enacted by Boeke's book, I argue that the animation process primarily en-gages its own techniques of mediation in order to obscure the seams be-tween distinct scalar milieus. Central to this process is the manipulation of temporal scale to fix scalar coordinates for the viewer. What emerges is a particular form of anthropocentric, nonlinear access to an underlying lin-earity—what I refer to as "equidistant optics," a new topology of knowledge production in which all scales are medially equidistant from the human knowledge producer. This mediated experience invokes Einsteinian relativity in order to exchange time for space, a scalar shell game that I analyze as a form of scalar collapse particular to cinema.

Chapter 4 continues my reconsideration of the filmic work and design philosophy of the Eameses, detailing the tortuous process of revising and remaking *Rough Sketch* into 1977's *Powers of Ten*. My focus in this chapter is on the relation between disciplined knowledge production and scalar mediation. In remaking their film, the Eameses altered its aesthetic and ideological framework in significant ways, producing a view of the cosmos in which underlying contiguity is broken up into distinct knowledge domains—physics, biology, geology, astrophysics, and so on, each a form of containment as well as a framework for discovery. Disciplinary access then produces intensifications within its constructed continuity, authorizing human access and mastery as normalized medial encounter. This arrangement perpetuates itself: because knowledge is scale-disciplined, it only ever concerns itself with a small slice of the scalar spectrum, and thus the illusion of scalar contiguity is maintained. Nonetheless, despite the mesmeric qualities of the final version of *Powers of Ten*, it still bears traces of the contradictory impulses that determined its construction. I trace some of these, including debates between Phylis and Philip Morrison, scale-obsessed science popularizers who became principal collaborators on the film, and the Eames team. I close this chapter by considering the political and aesthetic reversal that took place when the Morrisons, along with Ray Eames, remediated the 1977 film into a 1982 book for Scientific American. My underlying argument is that any instantiation of the cosmic zoom is essentially a form of ecology, an act of knowledge production that constructs a particular "shape" for the cosmos, as a networked constellation of scales. The cosmic zoom, then, is not merely a medial form but a framework for precharacterizing the scalar spectrum's differential potentials for encounter.

Chapter 5 constitutes the philosophical heart of this book. Here I fully develop the concept of *scalar difference* that has remained implicit up to this point. What kind of difference do we encounter across scales? Following early philosophical work on identity and difference by Gilles Deleuze, I distinguish between *primary scalar difference* and *secondary scalar difference*, arguing that this distinction is crucial for developing a non-anthropocentric understanding of scale. Primary scalar difference is an immanent form of intensity, enacted by matter-energy itself. This leads to the conclusion that scale is, ontologically, difference itself. Secondary scalar difference emerges comparatively after we have stabilized not only particular scalar milieus but objects within them. Secondary scalar difference is being across difference; primary scalar difference is becoming. Such a conception of scale, in turn, provides a framework for theorizing the interface between digital media and analog environments. I use the example of 1999's *Powers of Ten Interactive* to explore the affordances of applying a posthuman conception of scale to human-made medial systems, arguing that such systems have the potential not only to stabilize scales in new and more generative ways, but also to enable trans-scalar encounter as an embodied form of multiscalar ecology.

Chapter 6 extends this discussion of scale, mediation, and ecology to the multiscalar milieu of database-driven culture. In the age of big data and social media, scale and scaling operations are widely understood to constitute key dynamics and affordances of the digital. This chapter harnesses the long genealogy of the cosmic zoom and the theory of scalar difference developed throughout the book to recontextualize digital mediation as a trans-scalar force of composition rather than one of analog-to-digital mimesis. Beginning with digital cosmic zooms (including 1996's *Cosmic Voyage* and an interactive Flash applet) that harness the pixel and the database as twin affordances, and leading to recent cinematic examples that offer a counter-aesthetics of the cosmic zoom, I trace a shift in scalar mediation from an aesthetics of contiguity to one of multiscalar aggregation. Historicizing the functional affordances and constraints of the database, I trace its technical evolution from hierarchical and network organizational models to that of the contemporary relational model. I argue that this shift inaugurates a scale-instability at the heart of the database form, leading to a concomitant medial shift from trans-scalar *access* to multiscalar *address*. At stake in this chapter is the status of the subject within the milieu of database-driven media. I explore the subject jointly produced by neoliberal capital and urban banality as defined by a scalar constellation of switching nodes, recursively traversing the same terrain from variously interior and exterior modes. From the scalar flattening of neoliberal capital to mass profiling and surveillance, identity is distributed differently in our database milieu than it has been in the past, largely due to the relational database's recursive mediation of scale. As database-driven media produce and store information at scales radically at odds with subjects' own self-narratives, and then address subjects using newly stabilized scalar coordinates, the classical subject faces a desperate crisis.

While there are many possible tactical responses to the subjugation inherent in database culture, I suggest the possibility of a cautiously affirmative, creative one. Chapter 6 therefore develops the notion of the *recursive self* to describe such a hypothetical subject, one that would embrace the ontogenetic, creative affordances of the database to enable a recursive encounter with oneself across scalar alterity. In this way, contemporary media can enable a potentially radical shift in scalar subjectivity, a new medial mutation in identity capable of following—as contour and as critique—the oppressive scaling operations of our contemporary sociopolitical order. The continued existence of many beings, including our own species, may depend upon the innovation, at the heart of our cultural practices, of a scale-fluid subjectivity that fully engages—rather than foreclosing and controlling—the radical potential of the trans-scalar encounter.

The cosmic zoom is, in one sense, a way to shore up the fragments of the human against our (scalar) ruins.[56] As I explore throughout this book, however, it also contains the basic provisions for an approach to identity and alterity that holds out the promise of saving, not "us," but the radically transformative potentials of the trans-scalar encounter.

surfaces of mediation

Cosmic View as Drama of Resolution

A SENSE OF SCALE

In July of 1956, Dutch schoolmaster Kees Boeke pitched a radical idea to New York publisher Richard Walsh: a book about scale. Titled "A Sense of Scale," it would impart exactly that to its readers.[1] The cultivation of a cosmic view, the ability to view one's surroundings as a set of nested scales, is, he argued, an urgent pedagogical task. In his foreword to the book, published the following year as *Cosmic View*, Boeke writes, "We tend to forget how vast are the ranges of existing reality which our eyes cannot directly see, and our attitudes may become narrow and provincial."[2] Where are we to look to transcend the narrowness of human perspective? For Kees and his wife, Beatrice Cadbury, we must first free our gaze from the self-reproducing ideologies and power structures of capitalism and majoritarian democracy, which together produce a hermetically sealed reference system, their own enclosed scalar provinces.

Beatrice, a daughter of the co-founder of the British firm Cadbury Brothers and thus a shareholder in the family's commodity empire, became radicalized in the 1910s, married Boeke, and gave away her company shares to the workers in the Cadbury factory.[3] This established a pattern of experimental, bottom-up management and radical communitarian decision-making that marked all of her and Boeke's future endeavors. Boeke had become active with the Quakers (of which the Cadburys were members) in London during World War I, drawn by their pacifist stance toward that conflict, but became most impressed by their anarchic governance procedures.[4] The Quakers make all decisions as large groups. Rather than vote, they convene a full group meeting, discuss the issue at hand, and do not move forward with a decision until and if consensus is reached. Kees and Beatrice adopted this basic principle and developed it into a radical education program committed to nonhierarchical power structures and bottom-up community building. The result, from 1926 onward, was the Werkplaats Children's Community (Werkplaats Kindergemeenschap), a primary school in the town of Bilthoven, outside of Utrecht. At Werkplaats, the students are referred to as "workers" and take on responsibility for maintaining, expanding, cleaning, and organizing the school, including growing vegetables and fruit, building furniture, and constructing educational devices.[5] All attend a weekly meeting during which they determine, by unanimous decision, what changes should be made to improve their educational community. This nonhierarchical system extends to the development of the curriculum, promoting a sense not only of shared responsibility but of creativity.

For Boeke, the goal—which he suggests should be the goal of all education—is to prepare students for "their bigger and more complex responsibilities as citizens of the world. This will require not only knowledge, but understanding, tolerance, and sympathy. It cannot be taught; it must be experienced."[6] The Werkplaats is designed not for the efficient transmission of knowledge but to provide the active experience of participatory world citizenship. The school was an experimental model meant to be scaled up to replace national governments. This process, which Boeke and Cadbury call "Sociocracy," was to begin locally, with neighborhood meetings in which local issues would be resolved. "We, ordinary people, must just learn to talk over our common interests and to reach agreement after quiet consideration, and this can be done best in the place where we live. Only after we have seen how difficult this is, and after, most probably, making many mistakes, will it be possible to set up meetings on a higher level."[7] Once this system was functioning properly, local leaders would emerge to represent the neighborhood in larger ward meetings, which would operate at a somewhat broader scale and in turn send representatives to district meetings. Eventually, "40 or 50" of these assemblies could care for "the whole population of a small country." But the process was not to stop there: district representatives would convene a central meeting, which would send representatives to a larger-scale meeting, and so on, until all of the earth's continents are represented at a single "World Meeting to govern and order the whole world," a process that would involve the "reasonable distribution of all raw materials and products, making them available for all mankind."[8]

This unitary, bottom-up, consensus-only world government would operate within a noncapitalist framework where competitive advantage is impossible. Such an arrangement would require not only the skills that Werkplaats is designed to teach through experiential immersion, but also a large-scale engagement with the raw "materials" of the earth, the technical systems that enable their extraction, decomposition, and distribution, and a rational system of recombination and redistribution to ensure equal consumption and constructive, noncompetitive use. Such was Beatrice's and Kees's dream of a radically egalitarian form of globalization—a dream that took the form of a communitarian cosmic zoom.

By the early 1950s, Kees Boeke had decided that the expanded, creativity- and responsibility-fueled worldview the couple hoped to develop at the Werkplaats required, as a necessary condition, an understanding of scalar alterity—a "cosmic view"—and thus embarked upon a collaboration with the students to speculatively explore other scales.[9] As part of the ongoing project, students were asked to draw their surroundings at ever larger and smaller scales, generating many of the first prototypes for the book. By 1956 these projects had, through feedback from scientific experts and meticulous refinement, coalesced into the final volume, illustrated by Els de Bouter, a resident of Bilthoven and a former student of the Werkplaats. *Cosmic View* was published first in English in 1957 and then in Dutch two years later. True to its origins, it presents a narrative of the Werkplaats. The first image is a photograph of a young female student seated in the school's courtyard with a large white cat on her lap (figure 2). The caption explains in detail that the image on the page, measuring 15 centimeters square, is exactly one-tenth the size of the original scene. It is a 1:10 scale image.

Boeke's text restates the scalar relationship between the size of the image on the page and the area being described in several ways, going so far as to provide both metric and imperial units. Why belabor this banal cartographic point about the seemingly obvious relationship in size between the page of the book and the original scene? I suggest it is a question of media materiality. The outcome of a long process of optical enlargement, photosensitive printing, press layout, and final printing, the size of the viewing medium (in this case a halftoned ink pattern on a sheet of paper bound in

Figure 2 ▸ The inaugural scale, page 6 of *Cosmic View*: 1.5 meters × 1.5 meters.

a codex) bears a precise relationship to both a material arrangement that once existed and a speculative arrangement within which we're enfolded by the text's narrative: a girl and a cat, ten times larger than this image, sit in a courtyard—what will happen next? The medium through which this narrative is expressed is not arbitrary and is not transparent (figure 3). The reader, holding the book in her hands, bears a determinate spatial relationship to the scene depicted therein.

Image number two: a drawing of the school's courtyard, its edges just now barely visible. Two cars are parked on one side, and half of a dead whale is visible at the other. The girl, in her chair and with her cat, is visible in the center of the frame. A square box, representing the edges of the previous image, has been drawn around her. The text explains that this small box is 1.5 centimeters square, ten times smaller than the previous illustration. This trace of the preceding image will recur for the next twenty-five images, a reminder of what we've just seen and its scalar relationship to

Figure 3 ▸ Enlarged one-inch-square section from page 6 of *Cosmic View*. Individual ink drops and imperfections in the paper's composition are visible at this scale (4:1).

the current image. Our full image is now at a scale of 1:100, or $1:10^2$ (ten to the second power). The whale, largest known mammal, bears a promise of future scalar exploration: "Later this will enable us to observe in how many of the domains of scale living creatures occur."[10] The entire whale's body is visible at the next scale, $1:10^3$, along with the full school and its grounds (figure 4). The shape of the central building is now intelligible; according to the text, "The U-shaped building . . . tells a story of the war years, when the German military built it during the occupation of the Netherlands. After the liberation it was rebuilt and enlarged to become the central building for the Werkplaats Children's Community."[11]

Something odd has happened here. Isn't the book's lesson supposed to be about scalar relationships and powers of ten? The scale of this building is clearly important; the relevance of its history is less obvious. No instantiation of the cosmic zoom over the next forty years will include such seemingly extraneous, nonscalar trivia. But if we read *Cosmic View* not as

Figure 4 ▸ Scale "3," in *Cosmic View*.

an exercise in mathematical manipulation of powers but instead as a self-reflexive, material mediation between the reader and a real assemblage of matter, then such details become relevant *and* articulated to scale. With a field of view 1.5 meters square (the initial photograph), the material arrangement documented by the image exists by virtue of the actions of at least three actors (the young girl, the photographer, and the cat), each of which has a history. In the second image, the appearance of the whale and cars suggests further histories (which the text takes pains to circumscribe, foreclosing inquiry into how the whale has come to be in the courtyard). These histories of movement actually matter; they are integral (even if implicit) to a full description of the material assemblage depicted. As the reader has been articulated to this assemblage through a definite scalar relationship mediated by the material conditions of the book itself, the question of what exists and why—that is, what is inside and outside of the assemblage—constitutes much of the drama of the narrative that is unfolding for the reader. The building in image 3 (figure 4) is revealed as a building only *at this scale*. For the reader, it exists only at a particular point in the narrative and from a particular scalar point of view.

The text's explication of the building's history amounts to a description of its existential conditions: World War II, the Nazi invasion and occupation of the Netherlands, wartime construction of buildings with protected courtyards, the post-occupation decision to reproduce the existing building in more durable form, and its purchase by a member of the Cadbury family for the purpose of founding a radical pedagogical institution.[12] This history *comes into view* from a narrative standpoint in order to explain the dynamics of which the reader is a part, but only once these dynamics (the relationship between the U-shaped building and what has come before) have emerged. This text could just as easily have been included in the author's foreword, where the relationship between the book and the school are explained. By withholding this information until the building comes into view, the text signals both that such historical information is only relevant to certain scales (here, when the entire building is visible and not yet so far away that the details of its construction cannot be made out by the reader) and that the material connection between the reader and the landscape thus imaged and described is rearticulated at every scale. Not only does the scalar relationship itself change (each page indicates, and thus produces, a different relationship) but the dynamics that become significant (i.e., required to explain the current state of the assemblage) are themselves scale-relative.

Every turn of the page presents the reader with a new scalar relationship between the material conditions of the codex and an indicated assemblage (girl and cat, U-shaped building, etc.), smuggling with it a set of historical dynamics that can be resolved at that scale: a student's ownership of a pet, the material history of the Nazi occupation.

RESOLUTION

Scale 4 delivers yet another novelty. "It is surprising," we are told, "that already in this fourth illustration the child, who filled the greater part of the first picture, has completely disappeared."[13] Indeed, we can barely make out the building that we explored in such historical detail one page earlier. "The reason," the text explains, is that each scalar jump shifts our vantage point ten time higher. As we are now looking down from a height of five thousand meters, higher than Mont Blanc, Europe's tallest mountain, it is "no wonder that the huge whale can now hardly be distinguished." However, something new comes into our field of view, something other than the surrounding fields and edges of a town. The text narrates for us: "We notice a strange wavy line reaching the school building. We wonder what that is. The next drawing will show."[14]

The text has eased us into its narrative gently. There exist a girl and a cat, sitting in a courtyard, bearing a particular scalar relationship with ourselves as readers, as mediated by this printed book. But as we (again, mediated by the book) expand our field of view, new dynamics come into play. Sometimes features come into view (like the wavy line) whose dynamics are not yet, at this scale, clear. But rest assured, they will be revealed when the proper scale is reached, the scale at which such dynamics make sense. At the same time, simultaneous to these emergences, the objects and dynamics we discovered at smaller scales are fading away, marked by a trace of the previous scalar boundaries in the form of a box. The features within can barely be discerned, then disappear from view entirely. What this book has enfolded us within is a *drama of resolution*. Jumping scales here means simultaneously resolving new details—perceiving and understanding dynamics that were not visible previously—and losing sight of other features as they fall below our resolution threshold.

"Resolution," a word that has seen renewed duty in the age of electronic media, derives from Anglo-Norman and Middle French *resolucion* and has two meanings dating to the thirteenth and fourteenth centuries: (1) "reduction or separation of a material object into its component parts or elements" and (2) "return of a swollen or inflamed organ or part of the body to its normal state."[15] In its current English form, the first meaning has been adapted into the optical property of differential detail, while the second has been abstracted from the somatic and adapted to the arts of music and narrative. "Resolution," then, signals both the capacity of a medial system to differentiate between changes in appearance (in, for example, a surface being imaged, or a screen displaying graphical information) and the correcting of a disharmonious balance in sound or narrative structure to regain (for the listener or reader) a sense of comfort, normalcy, and satisfaction. It is in the former sense that we clamor for ever more megapixels in our gadgets, and in the latter that we breathe a sigh of relief after our latest purchase, knowing we have resolved the gnawing discrepancy between the capacities of our cameras and phones and the capacities of those advertised online. Of course, such resolution is only temporary; the drama of appendage inflammation will continue as an endless serial, whether it is a case of agitated kidneys or obsolete iPhones.

After the invention of the telescope revealed that astronomical phenomena could be optically "resolved" into smaller component parts (the vague glow of the Milky Way was actually composed of many smaller stars, for instance), astronomers began to speak of the "resolution" of optical devices: the amount of detail they could reveal, the smallest difference they could distinguish. In both the astronomical sense and the case I've been discussing—images printed in ink on paper—resolution is a function of two material assemblages. The first is the display medium: the number of physical pixels in a screen, the number of ink spots a printing apparatus is capable of positioning on a given type of paper.[16] The second material assemblage that enters into a resolving relationship is that which is imaged or referenced on the first: clusters of stars, patterns of interconnected atoms, and so on. More formally, resolution as procedure—the resolving of detail—produces a set of differential relationships between these two domains, each of which functions within this relationship as a surface: the surface of a page or screen places itself in a referential relationship with some other surface at a different scale.[17]

The notion of *surface* that I am utilizing here refers simply to a contiguous spatial assemblage. All mediating technologies present themselves as surfaces in this regard, whether two-dimensional (the classical surfaces of the codex page or cinema screen) or three-dimensional (sculpture, architecture, virtual reality). These medial contiguities then reveal or produce a second, referenced surface. This latter, which I have described here as either a scene or an assemblage, must similarly be rendered as a contiguous whole intelligible in bounded space. Even the most three-dimensional, topologically complex, or abstract assemblage must, to be mapped or represented on a medial surface, conform to its affordances and itself become a surface. Rather than considering volumes and surfaces as mutually exclusive categories, then, I am suggesting that in the context of scalar media, volumes can be resolved as surfaces without negating their three-dimensional nature. The point is topological: when a volume is rendered as a surface through medial negotiation (the placing of one surface into a determinate relation with another), it must be arrayed on a plane. This may imply a two-dimensional flattening, as in a photograph, or retain a three-dimensional modality, as in virtual reality; either way, media representation on a plane does not imply the erasure of depth encoding, as the example of elevation contour lines in cartography demonstrated in the previous chapter. Whether and how dimensionality is visually flattened is of secondary concern; ultimately, scalar media establishes a *conceptual* plane.

When two or more planar surfaces meet within a medial platform, determinate correspondences and differences can be established across scales. A pixel changes color or a dot is placed at a particular position as a negotiation of one possible relationship between two surfaces, constrained by the capacities of each to manifest difference. Put more simply, resolution is a negotiation of difference. This negotiation, when narrativized, is what I am referring to as the drama of resolution.

AGENCY AND THE NEGOTIATION OF DIFFERENCE

Because any act of resolution requires a reference surface and a medial surface (either of which may be actual or virtual), its constituent negotiation of difference may involve a spatial dimension and a temporal dimension. Spatially, topographical and topological difference is resolved, or stabilized, through a process of bringing one system into correspondence with another system. For this purpose, spatial boundaries must be imposed on both: the edges of a screen, the boundaries of an eventful surface. This stabilization of boundaries is not external to the medial system but integral to it; its rules, protocols, limitations, and affordances emerge out of the shared possibility space of the two surfaces. There are three simultaneous and entangled processes, then: (1) the extraction, from a shared possibility space, of a dynamic of differentiation; (2) the stabilization of two distinct surfaces; and (3) the stabilization of edges or boundaries around each surface.

In *Cosmic View* (1) consists of a distillation of the cosmos into forty isolated "jumps"; (2) consists of the discrete pages of the book, the context-imbuing text that explores the scene of the drama unfolding, and the realist conventions of photography and cartography that evoke a second surface to be captured; and (3) consists of individual image frames and conventional units of length, in multiples of ten, that determine the boundaries of the reference surface.

We must not assume that any of these exist in advance of the process of negotiation that stabilizes them. This is not to say that they are not real, but simply that their specific properties emerge only through the making of what Karen Barad calls an "agential cut."[18] This follows from the nature of quantum mechanics, as elaborated by Niels Bohr: matter does not possess determinate characteristics until an apparatus of measurement, such as the one we are describing here, differentiates one part of matter from another, thus giving description to a concrete set of properties. "In other words," writes Barad, "relata do not preexist relations; rather, relata-within-phenomena emerge through specific intra-actions. Crucially, then, intra-actions enact *agential separability*—the condition of *exteriority-within-phenomena*."[19] Difference is not a function of the separation of object and subject, a condition of objective exteriority; rather, difference is immanent to matter itself and arises processually rather than categorically. Barad's point is a dual one: first, that matter isn't *a priori* divided into determinate objects; second, that its subsequent division is enacted not by external observers but from within, by the very matter being differentiated. Matter differentiates itself, and thus meaning arises as a material rather than merely ideational process: "Intelligibility is an ontological performance of the world in its ongoing articulation."[20] We will return to these points, fundamentally Spinozan in spirit, further on. What is significant to note here is that a mechanism of resolution, through a process of negotiation, produces a set of determinate properties for and between two surfaces that are nonetheless fully real. The process of resolution is thus fully material and fully discursive at the same time, belying the claims of those who view objects as preexisting their active determination through an agential cut, as well as those who would collapse ontology into epistemology with the assertion that "there is nothing outside of the text."[21] I will proceed with the assumption that matter precedes its interpretation but also hew closely to the insights of quantum mechanics, which tells us that objects as such come to have determinate properties and borders only through acts of differentiation (immanent to matter itself) that utilize mediating mechanisms to determine those properties and borders (for Bohr, apparatuses of measurement). For our purposes, we need not insist on these ontological details at this time. Our object is scalar mediation, and a close examination of the film *Rough Sketch* will, in the next chapter, furnish us with an excellent model for the entanglement of matter and discourse, ontology and epistemology. The key point for the moment is that *mediation* in the sense I've been developing is a primary process by which matter is differentiated into entangled yet distinct surfaces.

At a fundamental level, media can be understood as differentiation, an immanent mechanism that makes agential cuts. This apparatus is plural because it differentiates itself, becomes two or more surfaces that stabilize a material and semiotic relationship between themselves. We are already in the territory of posthuman media theory, which views media not as communication or representation but as a fundamentally generative process, an ontological act of ongoing creation rather than a mimetic mode of transmission. Matthew Fuller, working from Nietzsche's insight that the human subject arises only as an effect of the ontological interplay of forces (not as their cause, as we commonly like to think), develops the concept of an "inherent medial will to power."[22] Media, in this conception, are material systems that constantly differentiate, evolve, and recompose themselves, forming ever-new configurations and potentials through internal differentiation and structural coupling with other assemblages. According to this understanding, a medium is not a dissemination apparatus that can be used to send any message—it is not a communication channel. As Marshall McLuhan elegantly argued, the medium *is* the message.[23] This extends to media the capacities of matter that Deleuze and Guattari describe as "machinic," which is to say, autonomy, heterogenesis, productive synthesis, desire for or drive to connection: "Production as process overtakes all idealistic categories and constitutes a cycle whose relationship to desire is that of an immanent principle."[24] As Manuel De Landa has shown, over time these processes of differential production generate large-scale flows and concretizations of matter at the historical scale. In other words, they produce ever-larger structures, which, once solidified, generate further divisions and differentiations at a whole range of scales: "The structures generated by matter-energy flows, once in place, react back on those flows either to inhibit them or further intensify them."[25] Difference generates its own immanent intensification across both spatial and temporal scales, from the dance of quarks to the death of galaxies.

Media, including familiar, seemingly human-scale media, should therefore be understood as a series of machinic differential operations within this cosmic-historical flow of matter-energy. In this sense it is apt to describe meso-scale media as, at the very least, infrastructural. As Nicole Starosielski has argued, "Analyses of twenty-first-century media culture have been characterized by a cultural imagination of dematerialization: immaterial information flows appear to make the environments they extend through fluid and matter less."[26] The cure for this topological mania will no doubt require not only a robust "network archaeology" of the sort Starosielski practices in relation to material network infrastructure, but also an explicitly materialist account of the ways that such infrastructures produce new assemblages at scales beyond their originary topographical coordinates—in other words, a scale-informed analysis of intertwined topological and topographical dynamics. At very large spatial and temporal scales, it is also apt to describe media as "geological" and treat them both as layered strata in deep time and as fundamentally extractive forces sustained by the mineral potentials of the earth, as Jussi Parikka emphasizes.[27]

For a moment, let us molecularize and schematize this medial process, extracting its basic operational diagram. First, there is matter, endlessly differentiated and differentiating. Media emerges: one medium (surface) engages with another. The two are asymmetrical, differently constituted. A negotiation must take place, such that their differences are resolved (but not *dissolved*) into a determinate set of relationships. In the advanced case of *Cosmic View*, a chocolate company, Marxist idealism, the Nazi war machine, and a group of schoolchildren, *inter alia*, eventually effect a sufficient set of entangled differentiations to give rise to a crystallization in the flow of matters and concepts: a book that utilizes the mediums of paper and ink to place a series of dots in a square, arranged in a pattern such that each page bears a determinate relationship to an external surface of a certain size. The printing press, ink, and paper together possess a resolving capacity of approximately 260 dots per inch, according to my count (figure 3, above). The first surface resolved in the book possesses its own differentiations: a child, a cat, and a concrete slab. The stabilized relationship marks its boundaries at 1.5 meters by 1.5 meters on the concrete slab and 15 centimeters by 15 centimeters on the page. This medium affords only one binary difference: white paper fibers or black ink drops. Here, the number of differentiated details of the reference surface is constrained by a medium (the medial surface) capable of resolving approximately 2.3 million binary points in image form as well as some 2,000 alphanumeric characters (which are also images, of course) in textual form.

The result of this medial negotiation between the differential characteristics of two surfaces is that a certain amount of detail (differentiation) of the reference surface is resolved as a page in *Cosmic View*. A different medium would resolve more or less detail. It is always theoretically possible to increase or decrease resolution. But for any surface's given capacity to differentiate detail, only a limited "picture" of another surface can emerge, subject to an absolute limit given the capacities of the resolving medium. We have now arrived at an articulation of scale in medial terms. Given two different surfaces that stabilize a determinate relationship between themselves, constrained by the one with the lowest capacity to register difference, scale emerges as the limited set of features that such a system can differentiate. I will offer, then, a provisional, very general definition: a "scale" is *a singular resolution of ontological difference between two surfaces*.

DISCONTINUITY AND MILIEU

Many implications follow from the definition of scale given above, and it will require the remainder of this volume to flesh out even a handful of them. To name a few, in skeletal form: the materiality of scale, the arbitrariness of scalar boundaries (there must be boundaries, but where they land is a function of a given mediating system), and the irreducible tradeoff between difference and boundaries—the larger the boundaries, the fewer (and larger) the differences that can be resolved. I have rendered this articulation of scale in terms of mediation, but I do not wish to imply that scale arises only as a result of deliberate *technical* intervention. We will explore many examples of scalar mediation and difference that are prior and resistant to the processes of the human, from the scale of the nano to that of the cosmic. Similarly, as this chapter attests, scales are mediated through discourse as much as by technology. In fact, the two are often inseparable, as analysis of the many instantiations of the cosmic zoom readily reveals. As we shall see, the relationship between techné and scalar mediation is complicated and problematized by these considerations.

Because scalar mediation not only preserves but helps to generate differences between surfaces, it always internalizes discontinuity. That is, the two surfaces whose differences are negotiated in an act of mediation are plucked, as it were, from discontinuous points along the scalar spectrum. This is precisely what Boeke highlights by prominently including a scalar ratio on each page of *Cosmic View* (reinforced by textual reminders that the page before the reader has its provenance at one point of the spectrum, while its referent surface hails from quite another). The scalar mediation performed by the book brings these discontinuities into temporary correspondence, enabling a form of trans-scalar encounter. The analog qualities of the book's paper fibers and ink flows cannot bridge the scalar difference between these two surfaces. In the sense that I introduced in the previous chapter, then, *Cosmic View* is analog media but employs a digital model of scale.

An equally salient implication of this particular theoretical understanding of scale as medially negotiated difference, however, is that a given scale that has been stabilized (resolved) through an act of mediation is necessarily a milieu. That is, it describes a spatial region that possesses certain resolvable characteristics. These differences are necessarily interrelated, forming dynamics and thus interdependencies. They are, then, what we think of as an *ecosystem*. Ecologists have long understood that any defined ecosystem is scale-relative. Different subfields of the ecological sciences take different scales as their resolution of choice, from extremely local (leaf, pond, estuarine) to "patch," community, and landscape ecologies. There has been much debate within the scientific discipline of ecology about the applicability of these scales and what may be "falling through the cracks" of this scalar disciplinization.[28] For now, suffice it to say that a given resolution implies a certain scale, and each scale implies an ecology. When Boeke reveals, at scale 5, the source of the wavy line introduced at scale 4, it serves as a reminder that the resolvable features of any scale are interdependent: "We see Bilthoven as a suburb of Utrecht. A dotted wavy line symbolizes a radio wave of 298 meters wave length reaching Bilthoven from the transmitter southwest of Utrecht, called 'Hilversum' after the town where its studios are."[29] A set of interdependencies arise that are resolvable only at this scale: Werkplaats is sustained by the town of Bilthoven, which receives information (at the very least) from Utrecht, its much larger neighbor. Utrecht contains, on its outskirts, Hilversum, which houses the infrastructure necessary to broadcast radio waves to Bilthoven. This infrastructural view revealed by Boeke's cosmic jump thus renders visible what Starosielski describes as infrastructures that remain culturally invisible by default. Media consumers depend entirely upon these material infrastructures for the consumption of their visible media, "but have little recognition of the structures of dependency into which they are often locked."[30]

The radio wave illustrated here, which is a cultural-material hybrid that links communities, is resolvable only at this single point along the scalar spectrum, where its origin and destination are both contained in a single milieu. A "milieu" in this sense is the subset of a total environment that bears meaning for a given observer. I am here following Jakob von Uexküll's concept of an *umwelt*, or lifeworld for an organism. Different organisms that occupy the same environment may inhabit very different *umwelten*, as they will have access to and uses for different features or details of that environment.[31] Boeke illustrated the Hilversum-Bilthoven radio transmission as a wave at the smaller scale, but it did not properly belong to that milieu, and the text quite rightly declined to identify it—its function there was as resolution-anticipation, one of Boeke's strategies for uniting scales within a speculative ecology within the reader's mind. This is intuitively correct: a radio wave cannot be properly "received" without a device with the correct protocols to tune in its wavelength, listeners who can decode the linguistic or musical information contained in the signal, and the cultural protocols to understand this information as the product of social flows that have a transmission point in Utrecht (or beyond). Thus while these interlocking systems occupy many scales, they are "composed" as a milieu only at a scale large enough to contain them all within a single *umwelt*—yet not so large that radio infrastructure can no longer be resolved by a milieu's inhabitants.

This is not to deny that a listener may, like a reader, jump scales while consuming media, concentrating on the effect of the sound waves on one's body at one moment and considering the origin of the signal at another. The point here is that different dynamics are available at different scales. At a scale of 10,000:1 (10^{-4} meters), the spatial and social relationship between Utrecht and Bilthoven is completely irrelevant to the dynamics we resolve. At scale −4 (figure 5), *Cosmic View* shows us a colon bacillus being attacked

Figure 5 ▸ Bacteria form part of a speculative ecosystem at scale "–4" in *Cosmic View*.

by bacteriophages, a virus, and "some gold leaf left on the girl's hand, possibly from book binding."[32] The book binding is a remainder, a remnant of the book's trans-scalar narrative. It has relevance only in our memory of another scale, far above and in the past. Now, here, within a milieu only 225 square microns in area, bacteria are concerned only with their survival in their battle with bacteriophages. The gold leaf, of unknown and irrelevant origin from their perspective, could potentially cut one of them in half (as the text explains), but this potential action would be quite unlikely in the ordinary course of life of these beings—what insurance lawyers refer to as an "act of God." God is agency that remains unresolved across scalar difference.

Boeke excels at invoking the drama of such scale-bounded dynamics. Part of his project is undoubtedly to demonstrate that *a lot is happening* at scales we, as bodies in the world, are intimately (inter)connected with but very infrequently resolve. All of the scales are interconnected, of course:

the girl's skin is the surface that provides the material ground for the vibrant bacterial ecosystem at 10^{-4}, just as the motion of the oxygen and nitrogen molecules that constitute air, at 10^{-7}, make aerobic life possible at larger scales and the Milky Way galaxy, at 10^{22}, contains all of the smaller-scale milieus explored in *Cosmic View*. The scales Boeke chooses are of course arbitrary. It is remarkable, then, that we find so many of these scales acting as milieus; that is, fully functioning environments for dynamic processes unconcerned with features outside of the immediate scalar environment described by a single page. There is drama—in the sense that all dynamic movement and self-assembly, all becomings, involve drama—at nearly every scale. And yet, a second-order set of dynamics emerges when we consider more than one scale together; that is, when a medial system affords the serial resolution of multiple scales. This process, at least in *Cosmic View*, couples the narrative, aesthetic, and ecological continuity of its analog form with the discontinuous (digital) scalar distance between juxtaposed surfaces that encounter one another. The rhythms, traces, and dynamics of this medial scale jumping are what I referred to earlier as the "drama of resolution." We can now, at last, more fully describe that drama.

THE DRAMA OF RESOLUTION

As we turn the pages of *Cosmic View*, we are led to engage with, at first, ever-larger scales. The size of the page doesn't change, nor does the page's resolution, which is a material constant. But while one surface (the page) remains fixed, the second surface, with which it is coupled, is systematically redefined. Its boundaries are expanded, and thus, in each successive image, a new relationship is established between the two surfaces. Without changing resolution, we have resolved a new scale: a courtyard, a city, the planet Earth, and later, as we shift to shrinking scales, a microorganism, a cell, the nucleus of an atom. At the same time, however, the medial system itself has expanded, in time and potentially in space, to incorporate multiple scales, and thus a third-order set of relationships emerges: relationships *between* scales for any observer articulated to (and thus part of) the medial system itself. In *Cosmic View*, each image after the first on the "outward journey" contains a box, 15 millimeters square, that delineates the boundaries of the previously rendered scale, and an even smaller box, 1.5 millimeters square, that delineates the boundaries of the scale rendered two pages prior.[33] We can think of these as traces, scalar retentions. As readers we've already experienced those scales, engaged their dynamics. Because they have been rendered smaller, without a change in page resolution, they contain decreased detail. Small differences in the surface being imaged are no longer discernible; only the largest of differences (such as that between the Milky Way galaxy and the blackness of the space that surrounds it) remain resolvable. Just one page earlier we could make out individual stars. They now exist only as a memory, though we are looking at the very same surface. To make one of the book's titular scalar "jumps," then, is to participate in a dynamic of image and memory, a change of detail, a play of difference. To view the same surface and yet see different difference: this is the drama of resolution.

As we jump higher and higher in *Cosmic View*, revealing ever-larger scales, details that we had previously resolved disappear from view. Early on the text emphasizes the moment when our original protagonists, the girl and the cat, can no longer be resolved on the page: "It is surprising that already in this fourth illustration the child . . . has completely disappeared." Just two pages later, when our scale includes a large portion of the Netherlands and Utrecht has been reduced to a dark smear, the text highlights features of our new milieu, including The Hague, Rotterdam, the Rhine, the coastline, but not before observing, "There is Bilthoven, and . . . there is the little girl: we know she must be there, but we cannot see her!"[34] Resolving new scales does not cause features at smaller scales to magically pop out of existence. Smaller and larger scales are *there* whether we can resolve them or not.

Roland Barthes, in his study of photography, suggests that this existential declaration—"*there is* . . ."—is the essence of the medium: "the Photograph is never anything but an antiphon of 'Look,' 'See,' 'Here it is'; it points a finger at certain *vis-à-vis*, and cannot escape this pure deictic language. . . . It is as if the Photograph always carries its referent with itself."[35] Boeke's text here seems to indicate, in exactly this pictorial manner, that the girl is there *in the image* even though we can't see her! What scalar magic is this? The girl, who has disappeared from view, has not disappeared from our consciousness. It is memory and text (and what is text but externalized memory, virtual memory?) that keep her alive. In Boeke's dramatic economy, the pleasure of the text is, at least in part, to watch objects slowly disappear while holding them *right there* in the center of the picture. This reflects the scalar dynamics of jumping to ever-larger scales: as more of a surface is imaged, less detail per unit can be resolved. But it has been recorded on another surface, the nervous system, as a memory of difference. The outward journey of a cosmic zoom therefore requires a continual increase in the density of remembered detail, what we might call the resolution of memory.

This play between what we know and what we see is fully exploited by Boeke for dramatic effect. At 10^{17} the image consists only of a single white spot amid the large black void of space. The text coyly describes this image as "very uninteresting," but then delivers the dramatic payoff:

> That spot, however, stands for the whole solar system, which on this scale would be only little more than 0.1 millimeter in diameter. In reality this illustration therefore is a very interesting one, because we now know and understand that that little speck of light contains not only the sun, but with it, all the planets, comets, asteroids, and meteorites which move around it . . . and their orbits! And we now realize that it is quite possible that numberless other stars that we see at night may have such satellites moving around them.[36]

The sixteen preceding scales we have jumped through to get this far have not only taught us about the dynamics we can resolve at each; they have progressively inscribed patterns of difference on the nervous system of the viewer that remain actively part of not only the surface being imaged (as in, for instance, a novel, which requires the reader to remember the plot that has transpired thus far), but also the surface of resolution: the book itself. Something remarkable happens, then, when the page before us can resolve only a single dot, and yet we "now know and understand" that it contains within it everything from a girl with a cat to the solar system we live in. One dot on a page contains, in this extended medial system that includes a third surface, the nervous system, just so much difference. The text asks us to consider any of the myriad stars we can resolve with the naked eye. This point of light, like the dot on the page, contains no difference except a binary one: it is not a void. It exists, it is there, nothing more. But when we train our travel-tempered, detail-laden gaze upon that point of light, when we are able to see it from a trans-scalar perspective, an extra dimension of differentiation is transversely added to our act of resolution. Here, we exceed the resolution of the eye, just as *Cosmic View* has trained us to exceed the resolution of the page. This, I argue, is what Boeke means by a "cosmic view," a "sense of scale." It is an understanding of the limits of resolution even as it is simultaneously a technique for apprehending and comprehending beyond the visual field.

AN INWARD JOURNEY

The second half of *Cosmic View*, a series of jumps to ever-smaller scales, starting from the same photograph of the girl in the courtyard, involves a similar process of accruing differential detail, but here the detail that exceeds, through time, the resolution of the viewing medium is no longer "in" the image but above it, behind it, after it. In some ways, then, the inward journey, through scales less familiar to most readers, is nonetheless more graphically intuitive. It is a process of magnification: new details are resolved, while the field of view is narrowed. The same tradeoff of resolution we encountered in the outward journey, inverted. At 10^{-6}, however, where we begin to resolve individual molecules of air, the image changes. The text explains that we have no imaging device capable of such resolution; "a more schematic image is therefore unavoidable, that is, a diagram rather than a photographic enlargement."[37] I shall have more to say about this in later chapters. For now, it is sufficient to note that the inward journey quickly takes us beyond the domain of the optical analog—what we resolve on our page hereafter bears no relationship with how the surface looks to a hypothetically tiny optical device. The images, no longer rendered as magnified versions of what has come before, will be diagrams meant to capture the essential dynamics that can be resolved (differentiated) at a given scale. At the time the book was written, neither the scanning tunneling microscope (STM) nor the atomic force microscope (AFM) had yet been invented. The problem, however, is not (as Boeke's ambiguous text could be interpreted to suggest) that we lack devices capable of producing optical images at these scales, but that the concept of the optical image cannot apply to them. The STMs and AFMs of later eras produce images through algorithmic processes that convert voltage differences at the atomic level into photorealistic images via a coded interpretative process.[38] Thus, even the scientific atomic imaging of today is diagrammatic in this sense.

As *Cosmic View* jumps to ever-smaller scales, it takes us deeper and deeper inside the student's body through a series of milieu diagrams. Within each milieu new relationships are revealed: relations of interdependence, of predation, and of production. At each scale, a new ecology. Our challenge, as readers, is to understand each new set of scalar dynamics not just on its own terms, as a resolved milieu, but also as part of a larger whole of irreducible scalar parts. Each scale contains interconnected entities but also houses processes tendriled with other scales.

In both the outward and inward journeys, *Cosmic View* taxes us not only with the burden of registering difference that disappears, but with fundamental shifts in the modality of our media assemblage that defy integration even as they call out for it. The girl is still in the picture, even if we can't see her. We must be trained to see her even when we are looking at DNA strands. In his introduction, Boeke implies that we do not ordinarily receive such training:

At school we are introduced to many different spheres of existence, but they are often not connected with each other, so that we are in danger of collecting a large number of images without realizing that they all join together in one great whole. It is therefore important in our education to find the means of developing a wider and more connected view of our world and a truly cosmic view of the universe and our place in it.[39]

As we shall see, latter-day practitioners of the cosmic zoom have often sought to splice disparate images into "one great whole," but it is clear from our discussion thus far that such a unity is not the simple outcome of achieving a wide-enough vantage point (or a medium with resolving power sufficient to image all of matter at once). Such a vantage point would be outside of matter, and thus could not be a component or product of a medial system that articulates one assemblage to another. Rather, achieving this "wider and more connected" view must, in Boeke's terms, consist of a rigorous *process* of resolving difference, a process that does not annihilate or assimilate difference into homogeneity, but preserves it in all its multidimensional, multimodal, multiscalar splendor, relying upon the surface of the nervous system as scalar memory to unify discrete experiences along the scalar spectrum into, not a continuous whole, but rather a tapestry of accreted discontinuity. *Cosmic View* is, itself, a diagram for resolving difference. A diagram of Sociocracy's aspirational subjectivity. In such a scale-aware society, resolving cuts that isolate particular scales as living milieus would impel us to negotiate relationships in and as difference.

At the scale of a global society, a medial cosmic view of this sort implies a third form of resolution, after that of material mediation and negotiated difference between scales: that of conflict resolution. Boeke, the pacifist, wanted to see an end to war, hate, and exploitation. For him, understanding the drama of resolution and scale is fundamental to a diffusion of conflict. Scale literacy, a cosmic view, centers difference as the foundation of all mediation, and mediation as the foundation of our engagement with the world around us, our many milieus. The citizens of a Sociocratic world must understand our mediated milieus as multiscalar, engaged through informed and egalitarian use of resolving cuts.

To make such resolving cuts requires not only access to particular forms of mediation but also, as internal provision, the catalyzing sense of wonder that leads one to attempt these jumps in the first place. Thus, while the book's outward jumps, to larger scales, help the reader to resolve the contextual (both spacial-territorial and temporal-narrative) differences necessary to form Sociocracy's ever-larger representative political bodies, its descending scales diagram the forces that compose all larger assemblages.

The exercise is one not of technoscientific representation (singularity) but of diagrammatic (abstract) processes of composition. When Boeke's editor, Richard Walsh, asked him to change the book's title to "The Universe in 40 Jumps: From the Atom to Infinity," Boeke vociferously objected because, among other things, "Cosmic View" "expresses better the tremendous mystery it deals with, and which I want to put before the reader." The title should "make them wonder."[40]

There is a world of difference between knowledge and wonder. Resolving difference requires both. One must apprehend the differential detail resolved by scalar jumps and preserve it when it dissolves into the miasma of our enabling medium's limits of resolution. For Boeke, the key to retaining scalar difference while integrating it is to foreground the material relationships established by every resolving cut. Turning each page in *Cosmic View* stabilizes a new scalar relationship between the perceiving subject and the reference surface. A "cosmic view" is obtained through a mediating surface that effects this relationship. One of the chief dangers in composing "one great whole," as we shall see in subsequent chapters, is that the mediating surface can drop out of our drama of resolution, reifying the whole as an unmediated, homogeneous surface. Boeke's ideal subject, by contrast, always holds the resolution of difference in medial tension, rendering fluid the relation between a surface of a milieu, a surface of mediation, and a surface of retention—the trans-scalar memory of the cosmic citizen to come.

an analog universe

Mediating Scalar Temporality in the Eameses' Toy Films

MEDIAL BECOMING

Powers of Ten, the 1977 short film by Ray and Charles Eames, is perhaps the most recognizable and influential cosmic-zoom work of the twentieth century. As such, it has been widely discussed in many disciplines and is often evoked in relation to the visual culture of scale. It has generally been approached, however, as a fait accompli, as the perfect instantiation of the cosmic zoom, along with whatever the cosmic zoom is assumed to signify. My approach in this chapter and the next will focus on the technical, conceptual, and collaborative processes by which the Eameses' two filmic versions of this work were constructed. The history of the project is a contentious one, marked by many dead ends, fierce debates, extensive research, and significant innovation.

The approach to scale that I have been developing thus far, attending to media materiality and analyzing the circuit of scalar mediation as the milieu-building negotiation of difference between surfaces, demands that we consider not only the interfaces that stabilize certain scales for human access, exploitation, and encounter, but also how scalar media come to be in the first place. If scale is a particular negotiation of difference, then scalar media cannot be straightforward scale-making machines. On the contrary, they are themselves inextricably bound up in processes of negotiation. Scalar media make scales, but at the same time, scale makes scalar media. To fully parse the scalar negotiations immanent to their construction is not only to understand the processes of scalar mediation more intimately but to discover the potentials for encounter that were eliminated, occluded, or rendered dormant in their architecture. These serve as warnings, but also sometimes as recoverable medial potentials, lying latent and unseen.

This chapter charts the early efforts of artist Eva Szasz and designers Ray and Charles Eames during the 1960s to create an animated version of Kees Boeke's *Cosmic View*. The translation of Boeke's book into the medium of film was fraught with difficulties both material and conceptual. Chief among them was the question of time: in animating static views of scalar milieus, what relationship would be established between the viewer and the temporal rhythms of those milieus? After all, the scalar spectrum is as temporal as it is spatial. How to mediate timescales that are not merely outside of human view but outside of human experiential time? As film is a time-based medium, it offers both established techniques for temporal mediation of scale—chief among them the zoom—and pitfalls for the unwary filmmaker attempting to capture something of scalar alterity within this most anthropocentric of mediums.

Film is an analog form of media, meaning that it operates through continuously variable intensities (differences produced on a surface of photochemical emulsion). Yet it is also digital, introducing invariably quantized packets of information in the form of discrete frames. Analog and digital modalities have profound implications for both medial temporality—the timescales that media stabilize and are capable of synchronizing to—and the resolving of other-scaled milieus. How do analog or digital media delineate their boundaries, and what forms of differential detail can they render? In short, the question of the analog is here a question of differentiation, the degree to which difference is resolved or smoothed into continuous contours. As I document in this chapter, the Eames Office struggled with these questions when it produced its first fully realized cosmic-zoom film, *Rough Sketch*. The filmmakers navigated these pitfalls partially by following formulas they had devised in making previous (and then later) films about toys. I will argue that the aggregate effect of these films was to construct an *analog scalar spectrum* marked by continuity, epistemic access, and relativized temporality. Perhaps ironically, the analog universe that emerges from this construction is achieved *despite* the limitations of analog media, not as a result of their unique affordances. I will continue to explore this paradox in future chapters. Chapter 4, covering the making of *Powers of Ten*, along with the *Scientific American* book that remediated its imagery and surreptitiously revised its ecological codes, will focus on the ways that scale mediates knowledge production across disciplines. This chapter is about the interrelationships between continuity, temporality, and scale in the filmic medium.

ANIMATING SCALE

The reputation and influence of Kees Boeke's *Cosmic View* reached something of a zenith in North America in 1968, coinciding with the height of the US-Soviet space race. Perhaps never had the world seemed so far from Kees Boeke's and Beatrice Cadbury's dream of a universal cooperative, and yet so close to achieving the technical expertise sufficient to produce, if not a cosmic view, at least a planetary one. In the United States, NASA was gearing up to send a crewed spacecraft to the surface of the moon. Stewart Brand, who had started a campaign in 1965 to ask of the space agency, "Why haven't we seen a photo of the whole earth?" finally got his wish in November 1967, when NASA's ATS-III satellite produced the first photograph of the entire disc of the planet. Brand used this photograph on the cover of the iconic first issue of his *Whole Earth Catalog* the following year.

This inaugural issue of the *Whole Earth Catalog* featured, among a number of book reviews, a short writeup on *Cosmic View*, declaring it "one of the simplest, most thorough, inescapable mind blows ever printed. Your mind and you advance in and out through the universe . . . and you realize how magnitude-bound we've been."[1] Especially for individuals like Brand, working toward such large-scale goals as world peace, material abundance, and ecological sustainability, the concept of imaging the earth as a single entity took on great currency and urgency.[2] It was in this climate that two production teams decided, at roughly the same time, to create a film version of *Cosmic View*. In Canada, the National Film Board funded artist Eva Szasz to create an animated version. In the United States, Ray and Charles Eames directed the efforts of their world-famous design studio and staff toward the production of a proof-of-concept "sketch" that could be used to secure funding for a future version. The book had long been on their minds. In 1963 they had produced a test that found 240 frames, played back over ten seconds of the viewer's time, produced a pleasing transition between two photographic plates at exponentially different scales, creating the illusion of a smooth zoom.[3] As a test, the Eameses would sometimes assign new staff members at the Eames Office the task of working out how to produce the full experience of the book as a film. Judith Bronowski took the assignment particularly seriously, as we will see.[4]

Both Szasz and the Eameses had set out to adapt *Cosmic View* to the filmic medium, and initially borrowed its title. The Eames Office would change the title of its first effort to *A Rough Sketch of a Proposed Film Dealing with the Powers of Ten and the Relative Size of Things in the Universe* (generally shortened to *Rough Sketch*). Szasz changed her title to *Cosmic Zoom* to emphasize the shift from static page views (or "jumps," in Boeke's terminology) to a continuous traversal of scale afforded by the time-based medium of cinema.

How did Szasz and the Eameses achieve the illusion of a "cosmic zoom"? The two productions adopted a similar technical approach: a series of two-dimensional images would be photographed on an animation stand. Between successive exposures, each image would be moved in tiny increments closer to or further from the lens of the camera (which would be refocused), such that the image is enlarged or reduced within the frame (for the "inward" and "outward" journeys, respectively). When one image reaches the limits of its range (when its borders are about to appear within the film frame or it has become too blurry), it is swapped out for the next image, drawn or photographed to a different scale. The new image is placed in relation to the lens such that it appears to match the scale of the most recently photographed frame, a "reset" that once again affords the filmmakers the full range of motion of the animation stand. Each piece of artwork in turn goes through this "zooming" process. For a number of frames just before and after an artwork switch, the film within the camera is not advanced and is thus double exposed, creating a cross-fade that further smooths the transition between the images.

I describe this process in detail to emphasize its material limitations and affordances. A continuous cosmic zoom assembled with an animation stand comprises (1) *image plates*, a number of two-dimensional artworks, each representing a discrete scale; (2) *camera moves*, a series of movements relative to the camera lens (the "zoom"); and (3) *cross-fades*, a series of double exposures that blend the details of one scale with that of its neighboring scale.

Szasz produced her image plates in the same way Boeke produced his: by drawing them. Each scale was represented as a hand-rendered color drawing (figure 6). The resolution of these drawings is a function of their physical size, the precision of the tools used, and the number of differential features Szasz was, within these limitations, capable of and interested

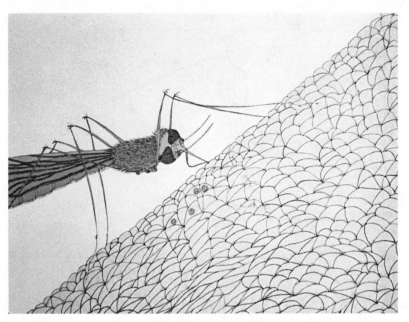

Figure 6 ▸ Frame from *Cosmic Zoom* by Eva Szasz (1968).

in including. The Eameses, aiming for a photorealistic product, utilized photographs for as many different scales as possible, resorting to drawings only at scales larger the earth and smaller than the cell. Szasz's film, *Cosmic Zoom*, begins with a live-action shot of a boy and a dog in a rowboat on a river near Ottawa. The shot freezes, transitions to cel animation, and zooms out to the edges of the universe, then in to the nucleus of an atom, and back out to the boy and the dog, who unfreeze and go about their nautical way. Throughout these scalar jumps, the camera gets closer and closer to one artwork before cross-fading to the next in a play of resolution: The 16-millimeter film stock can resolve only so much difference, clearly less than paper and ink can in each artwork. As the image plates move closer or farther from the lens, more or less detail is resolved at the film plane, some portion of which will be resolved by the film stock. Thus "zooming in" on an image plate reveals more detail (difference) up to the point where the cinematographic system's resolving power is higher than the amount of detail present in the portion of the artwork within its field of view.

In practice, then, *Cosmic Zoom* becomes a strange drama of resolution: At the beginning of each discrete scale (image plate) in the zooming-in phase, the differential detail of the artwork exceeds the resolution of the camera apparatus, generating a dramatic resolving of detail as it is enlarged. Soon, however, the level of detail available in the artwork falls below the re-

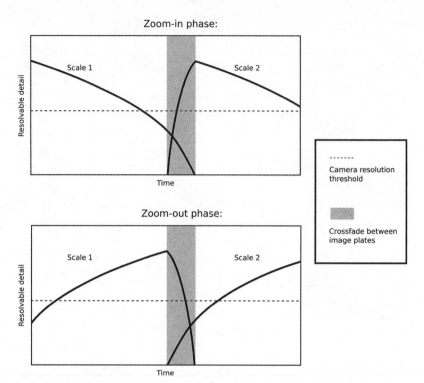

Zoom-in phase:

Scale 1

Scale 2

Resolvable detail

Time

Zoom-out phase:

Scale 1

Scale 2

Resolvable detail

Time

------- Camera resolution threshold

Crossfade between image plates

Figure 7 ▸ Arcs of resolution.

solving power of the cinematographic apparatus; for the viewer, the zoom continues, but less and less detail is resolved in the frame. Then comes the cross-fade, where the incoming scale (next image plate) has its dense detail superimposed over the forms of the outgoing image, now so impoverished in differential detail. Thus each scale remains a discrete milieu of detail, and though the field of view constantly narrows, the "seams" between scales are marked by dramatic changes in resolution over the short period of the fade. These seams register for the viewer as a sudden blurring of differential detail, a brief transition into a more homogeneous milieu. Figure 7 diagrams this relation between discrete scalar milieus, resolution, and time, and its mirroring in the zooming-out phase. While this describes the specific material process by which Szasz and *Cosmic Zoom's* filmmaking team created the illusion of a continuous zoom, it is the same technique used by the Eames Office to construct *Rough Sketch* and *Powers of Ten*. Indeed, I intend this to serve as a general diagram of the cosmic zoom's relationship with time.

The analog medium of celluloid animation consists of a series of surfaces: the surface of each image plate (drawn or photographed), the surface of the celluloid film frame, momentarily locked and pin-registered in the camera's gate, and eventually the surface of the screen on which it is projected (which we will get to shortly). Each surface is limited in its resolving power, or the density of differentiable detail it can contain. Further, each serves as an analog of some other surface: the image plate references a virtual surface elsewhere (as discussed in the previous chapter), the celluloid frame is an optically mediated analog of some part of the image plate, and so on. Producing a "zooming" animation requires that these surfaces encounter each other and transfer their detail from one to another, with transformations dictated by their medial properties and detail lost as determined by the relative resolving power of each. In this apparatus of production, image plates are analogs of scales, combined and converted into time-based media through the camera and celluloid. These make the cosmic zoom possible as a medial transformation, viewable on a screen by a spectator, but they cannot overcome, and must continually negotiate, the differences of detail bound up in all trans-scalar encounters.

The specific strategy of the cosmic zoom entails smoothing out the fundamental discontinuities between scalar milieus—regions of the scalar spectrum—by dynamically modulating changes of resolution between at least two of the surfaces entangled in a scalar relationship. As we will see, this time-based strategy of destabilizing the scalar negotiation between surfaces discussed in the previous chapter remediates the trans-scalar encounter into a frictionless form of access for the viewer. This is the cosmic zoom. The all-too-obvious seams, or ruptures, are the sites of medial reflexivity, reminding us that "scale" names both ontological difference and discursive negotiation.

ZOOM AND DISTANCE: "A ROUGH SKETCH"

The zoom lens is both an optical instrument and a modality. First patented in 1902 by Clile C. Allen, it has as its outstanding characteristic movable optical elements that allow for the adjustment of field of view (FOV) without changing either the focal point or the distance from lens to subject.[5] The camera operator can narrow or widen her FOV by substituting a tiny mechanical movement within her lens for a relatively larger movement of the camera apparatus, which (with a lens of fixed focal length) would need to be walked, driven, or slid forward or backward by relatively large units to achieve a similar change in FOV. But there's a scalar tradeoff: the effect of *zooming in* on a subject is distinctly different from that of moving closer. Long focal lengths "flatten" an image vis-à-vis the scene being photographed—because the camera apparatus is farther from the subject than it would be with a wider lens and the same FOV, the relative difference between distances to objects within the frame is lessened. This is why longer focal lengths are often preferred by portrait photographers: they flatten facial features, de-emphasizing "unpleasant" effects of perspective, such as elongated noses and sunken eyes, by altering the distance ratios between elements in the frame and the camera. With longer lenses, then, faraway objects can be resolved in greater detail through magnification, but only at the cost of losing information (difference) about relative distances to the camera.

The capacity for such perspective changes over a continuous period of time is a characteristic of the zoom lens exploited by Stanley Kubrick to great effect. In a signature zoom-out he employs frequently in both *A Clockwork Orange* (1972) and *Barry Lyndon* (1975), a faraway, flattened tableau is slowly revealed to be part of a much larger, far more dynamic milieu as the widening FOV admits the detail of depth into its frame. At the same time, such optical zoom-outs dynamically increase depth of field, increasing the planes of the image that are in focus and thus revealing more detail. The inverse is also true: zooming in reduces depth of field at the same time that it narrows FOV and flattens the image, allowing cinematographers to progressively isolate a subject from the maelstrom that surrounds it, a favorite technique of Sergio Leone, Robert Altman, and Sam Peckinpah during the 1970s—the golden age of the cinematographic zoom. Eva Szasz and the Eameses produced their cosmic zooms during the first wave of these techniques, imitating their aesthetics and drawing on audience expectations conditioned by mainstream cinema.

Yet the animation stand, as detailed above, does not actually zoom. Rather, it moves the camera closer or farther from the image plate. This leads Michael Golec to remark that the effect is more properly a crop than a zoom: "The film achieves its visual continuity of the frame through the technique of cropping, which is also the discontinuity at the center of *Rough Sketch*. To crop is to cut off or conceal parts of an existing scene or existing image."[6] In this regard, the crop, combined in each frame with an enlargement—the bringing closer of the image plate—functions more like a zoom than a camera movement. While camera movement alters the camera's relationship of distance to the various elements of its mise-en-scène, the zoom progressively erases such detail as it flattens the image plane. The cropping of the cosmic zoom uses camera movement to approximate the effect of the zoom because *the photographed subject has already been perfectly flattened by the process that rendered it an image plate.* The image plate is a two-dimensional surface. Thus while no zoom lens is involved,[7] Szasz's title is appropriate. At the same time, photographed artworks on an animation stand, unlike images photographed with a zooming lens, possess infinite depth of field, because all of their represented features occupy the same depth plane (the flat surface of the image plate), leading to an image that is always and completely in focus. The animated zooms of Szasz and the Eames Office, then, conjoin the flattening effect of the zoom lens with the unchanging depth of field of a fixed wide-angle lens. The effect produced in both *Cosmic Zoom* and *Rough Sketch*, with the exception of their opening shots (scales)—which are photographed in live action—is of a zoom that continually changes FOV without any corresponding change of depth of field or apparent distance to subject. In other words, it is a "zoom" that erases all referential cues to distance.[8]

This distanceless zoom places the viewer at an extreme remove from the unfolding scene. A paradigmatic example is on display during the films' opening sequences, in the discontinuity between the initial live-action shots, which seem to place the viewer somewhere within the same milieu as the human characters, albeit with a height advantage, and the subsequent animation-stand work that seems to remove the viewer from the scene entirely. Because the milieus that resolve themselves below us are effectively different and unique scales (each a play of difference on a two-dimensional surface), the cumulative effect of these cosmic zoom-outs and zoom-ins is to construct a position for the viewer that is outside of scale itself. Thus, while Boeke's *Cosmic View* maintained, with every page, a materially scalar relationship between the viewer (via the physical page of the book) and another surface at another scale, *Cosmic Zoom* and *Rough Sketch* retain a drama of resolution but disarticulate it from the scale of the viewer. Thus loosed from the material constraints of scale itself (which is always relational), the viewer is liberated to view scale as a cinematic menagerie: scalar difference can be conjured up before us, like a magic lantern show, while we remain outside and above such flickering phantasms.

It is all the more surprising, then, that the central motif introduced by the Eameses in their cinematic zoom is one of vertical travel. "We begin with a scene one meter wide which we view from just one meter away," intones the narrator, Judith Bronowski, at the beginning of the film, as we view a luxuriant picnic spread, complete with a sleeping man and, on the edge of the frame, a seated woman (played by Bronowski herself). This narrated distance cue establishes a neat unity between the default (original) scale of work and the distance of the virtual viewer from it, even before describing its features ("the sleeping man is having a picnic on a Miami

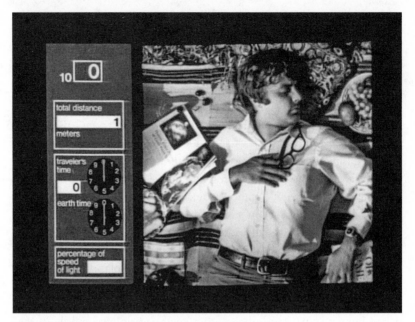

Figure 8 ▸ Frame from *Rough Sketch* by Charles and Ray Eames (1968).

golf course"). There is no mention here of the scale of the viewing surface (ordinarily a classroom projection screen); the viewer is assumed to be one meter above the scene, and thus the medial relationship between the viewer and the scene, so important to Boeke, is here replaced with distance as a stabilizing motif. As soon as the virtual camera, whose perspective is identified with that of the viewer and the instrument panel in our virtual spaceship, "backs off" (in the words of the narrator), the "total distance" begins to count up (figure 8). It eventually reaches 100 million light years before we dramatically zoom in, down to a distance of only 0.0001 angstroms from the nucleus of a carbon atom in the sleeping man's hand.

The presence of the instrument panel, along with the motif of traveling through space, replaces the medial negotiation of difference between surfaces in Boeke's original version with a fictional mediator: a vehicle that translates the resolvable differences of discrete scales into information for its passenger, a scale-fixed viewer who watches its instrument panel and "windshield" as multiscalar marvels are resolved thereupon. This motif borrows from popular science fiction films and television shows of the time, including *Fantastic Voyage* and *Star Trek*, both of which debuted in 1966, less than two years before work on *Rough Sketch* started. Equally, however, it arises out of NASA's Apollo program, which combined into a single cybernetic system a space capsule capable of transporting the human body as far away as the moon, the most complex instrumentation that had ever been developed to allow for the control of the craft, the largest single-purpose government funding mechanism with the largest number of employees since the Manhattan Project, and a system of media coverage that effectively universalized the (mediated) experiences of the Apollo astronauts through nonstop broadcasting. It was in this context that Russell Schweickart referred to his role in the Apollo 9 mission, during which he performed the world's first spacewalk with an autonomous life-support system, as "the sensing element for man."[9] In the same context the Eameses produced their film, which like Apollo begins on the ground in Florida and seeks to enable its viewers' virtual traversal of vast regions of space.

These features lead Sylvie Bissonnette to classify *Rough Sketch* and other filmic cosmic zooms as "scalar travel documentaries," motion pictures that take the viewer on a journey through scientifically derived scales yet "do not sufficiently inform viewers about the limits of the scientific models that they depict."[10] "Scalar travel documentary" certainly captures the discursive frame that *Rough Sketch* constructs for itself, and most scholars take at face value that the film presents us with a first-person journey through the cosmos, or at least a scientific model of it. I will take a different tack, however, and read the film as a form of scalar media that negates movement entirely. Rather than taking the viewer on a journey through anything, it *brings the cosmos to the viewer*. Another way of putting this is that rather than assuming a preconstituted viewing subject who it whisks away on a journey, the film constructs what I'm calling an *equidistant optics* that serves to stabilize (or construct) a particular subject by providing access to the full scalar spectrum as an enveloping medium. In this sense the film is something like a zoetrope, viewed from within as a magically animated universe spinning around the observer. I'll develop this model throughout the remainder of this chapter.

By harnessing the symbolic system of Apollo to evoke a vehicular interface indicating vertical spatial movement, yet graphically rendering that movement as a distanceless "zoom," *Rough Sketch* finds itself in paradoxical territory. Through this virtual interface, the viewer becomes a traveler who doesn't go anywhere, faced with an interface without a dynamic referent, experiencing a changing field of view without a corresponding set of dynamic milieus. Ideologically, then, the film's precondition of trans-scalar visuality, the assumed vantage point that makes its medial and conceptual project possible in the first place, is the universal perspective. The film begins visually at the human scale, but conceptually at the scale of the totality of the universe. The link between them is clear: the human possesses a universal gaze that renders the cosmic-zoom project legible to begin with. The default scale may be the human picnic, but the picnic's default perspective is universal. The political ideology of the zoom, in this context, gains its efficacy by providing the cosmic zoom's precondition: a totalizing view of all possible matter, all possible events—what Bissonnette refers to as "scopic mastery."[11] Where a fixed lens moving through an environment, as evoked by *Rough Sketch*'s narration, implies relative positionality within an embodied milieu, the zoom assumes a disembodied, mastering perspective that traverses detail without any direct encounter, without getting its feet wet. Geographer Denis Cosgrove argues that Western civilization has long sought such means of implementing an "Apollonian gaze," a transcendental point of observation from which all of life can be surveyed, and which therefore presents an illusion of mastery. "The imperial perspective is thus appropriately Apollonian: fixed in time yet mobile, globally central and divinely celestial." NASA's Apollo program fulfilled this dream of an Apollonian view of the planet, enabling us to collectively "marvel at a vast yet tiny earth."[12] *Rough Sketch* extends this multiscalar view further, speculatively imaging the entire universe as a sampling of scales down to the subatomic.

TEMPORALITY AND CONTINUUM

Most scholarship on *Rough Sketch* and *Powers of Ten* focuses on the dramatic continuity of their zooms, the aesthetic and conceptual smoothness they produce for the viewer. For instance, Sylvie Bissonnette argues that cosmic-zoom films are "insufficiently self-reflexive about the limits of their own representational practice, of scientific modeling, and their own production methods."[13] Michael Golec take the view that *Rough Sketch* is shot through with discontinuities, from its series of "broken images" to its breadcrumb trail of scientific references and extant storyboards to Bronowski's narration, a voice that "interrupts the flow of optical constancy and reveals the massive amount of information—the unseen masses—required to create the semblance of nondiscontinuity."[14] Essentially, the film is an open system that cannot help but encompass and point toward myriad discontinuities that haunt its presentationally closed system, its aesthetic continuum. I have been highlighting two entwined discontinuities: the material discontinuity of the image plates used to construct the film and

DOMAIN	Milieu	Medial system	Viewer
TEMPORALITY	Surface dynamics	Resolution of scalar difference	Medial unfolding
	Resolution change		

Figure 9 ▸ Divergent temporalities of filmic scalar mediation.

their attendant arcs of resolution that reveal the scalar seams of their stabilized milieus when animated in time. I now wish to consider a yet deeper form of scalar difference embedded in the heart of *Rough Sketch*, and indeed all scalar mediation: temporal scalar difference.

Figure 9 arrays four dimensions of time held in tension by the filmic apparatus. First, a surface's dynamics give rise to the timescale of interactions within each scalar milieu: at a given scale, how long do typical events take to unfold? Obviously I do not have in mind a mathematically derived number here, as a milieu's timescales are multiple and heterogeneous. Dutch scientists Gerard 't Hooft and Stefan Vandoren, following in Kees Boeke's footsteps, have produced a book that examines *timescales* in powers of ten. Just as Boeke demonstrated that most conceivable spatial scales are teeming with unique entities (there are a few exceptions), Hooft and Vandoren convincingly document events particular to a series of timescales, from the vibration of atoms to the decay of the longest-lived particles (which extend many times beyond the estimated age of the universe). "Every unit of time is unique. Every level showcases enthralling phenomena."[15] A surface's dynamics represent a cross-section of such timescales; which are relevant will depend upon the perspective of the entity occupying that milieu.

Second, within the domain of the medial system itself, the negotiation of scalar difference between surfaces—the resolving of scale (discussed in chapter 2)—has its own timescale. How long does it take to resolve (stabilize) a scale? This is tantamount to asking, "How long does it take to construct a medial system?" As I have discussed, in the case of *Cosmic View* it took many years and grew out of the establishment of the Werkplaats Children's Community, the co-production of a trans-scalar lesson plan, the production and refinement of many children's drawings, scientific research (which had itself developed over decades and centuries in order to achieve the capacity for microscopic and galactic measurement), negotiations with publishers, and the physical process of printing the book itself. The end result was the stabilization, for the reader, of forty scales. The process behind this stabilization unfolded over a span of decades or centuries (depending on how far we wish to extend our analysis).

Third, we must consider the temporality of the viewer, the final component of a medial system. Here we must differentiate between the time it takes for the entire medial system to unfold (for example, the time it takes to read Boeke's book or watch an Eames film) and the time required to resolve each new scale. As we shall see, this latter temporality is not necessarily nested within the former.

Finally, there is the question of resolution change. In *Cosmic View*, the text introduces a temporal assumption: for all intents and purposes, it assumes that its every mediation of scale takes place at the same exact instant: December 21, 1951, at noon.[16] All shadows (including the large-scale umbra of the earth), as well as the relative position of celestial bodies, are rendered accordingly. But what an assumption! Of course this means that the temporality of surface dynamics cannot be rendered graphically; as we have seen, however, Boeke utilizes textual description to indicate quite effectively the sorts of events and interactions that take place at each scale. Because his work is presented in codex form, the temporality of its medial unfolding is effectively infinite. The viewer can choose to scan each page at will. A change of page means a change of scale, but the interval of time between those page turns can be seconds or eons. Nonetheless, a paradox emerges when we consider the temporality of resolution change within *Cosmic View*. We may be tempted to say that this timescale is equal to the time required to turn each page.

But let us consider what turning a page entails. *Cosmic View* tells us that time is standing still. It equates distance with field of view, as does *Rough Sketch*, but as we have seen, it lacks the latter's motif of vertical travel (movement over time). In the absence of time associated with either the milieu's surface or the unfolding of the book, the viewer is assumed to have instantaneously jumped to a higher vantage point with each page turn. This seems more or less intuitive as we assume a virtual vantage point above Utrecht, then Western Europe, then the earth. We can imagine that different observers could have theoretically occupied these positions at noon of the appointed day, and that we now virtually jump between their perspectives. However, the deployment of this optico-spatial framework requires that we take into account the speed of light. In other words, vision is a function of distance and time. *Cosmic View* has given us the first two terms but denied us the third. Light cannot travel any distance without the passage of time, and thus it is not possible (even in a virtual simulation) to change viewing positions without the passage of time and still "see" the same events. In other words, we must move to different points in time as well as space if we are to view the same instant below.

We cannot hold steady both the temporality of a visualized surface and the temporality of the observer, and yet this is exactly what *Cosmic View* asks us to do. We may not notice the discrepancy at first, because the distances are close enough that we are used to disregarding light as a medium. But as our viewpoint assumes increasingly greater distances, and the text's fidelity to interactions and dynamics at different scales commits it to a description of the speed of light, a collision between the temporal constraints and the distance motif becomes unavoidable. At 10^{16}, when the solar system has been reduced to a tiny circle slightly over one millimeter in diameter, the text finally acknowledges the elephant in the room:

> As we have imagined all along that we are making our trip without spending time, this means that it would have taken the light rays which we now see more than six months to cover the enormous distance from the earth, even though they travel at the rate of 299,800 kilometers per second! It also means that if we had a marvelously good telescope and could see details of events on earth, the events we watched would be those that happened more than six months ago![17]

This text purports to describe the enormous distance of our viewpoint in relation to the speed of light, but it also contains a *mea culpa*. The "marvelously good telescope" is of course the virtual instrument of the book itself, whose conceit has required that the events described and depicted in earlier and later pages (during the inward journey) are simultaneous with events we could witness now were we to "unfreeze" time (the revolutions of the planets, asteroids flying through space, etc.). We have now come head to head with a temporal bifurcation that has been operative at every scale save one (the reference scale). If the surface we are observing has been frozen in time (the basic assumption of its graphical renderings), then the virtual observer must jump forward in time at every scalar jump. By this point, we would need to have jumped six months into the future, though the text has identified the timeframe of resolution change with the timeframe of the girl depicted in the first image. Alternatively, if we are assumed to have made instantaneous jumps to ever-greater heights, then the scene we survey has progressively moved backward in time, and what we view now took place six months ago. But in this case it is very unlikely that the girl and the cat have assumed the same position that they would six months in the future, and it is therefore no longer the case that they are directly below us and merely unresolvable. This bifurcation of time is itself unresolvable. We have discovered one of the basic implications of

Einsteinian relativity: there can be no such thing as simultaneity. For any system of scalar mediation, a change of resolution must be accounted for by varying the temporal relationships between milieus, the medial negotiation of scalar difference, and the unfolding of the medial system for the viewer. Another way to state this: at least one of these domains must temporally "absorb" the resolution (scale) change. There's no such thing as a free scale jump!

Rough Sketch sets out to rectify this problem by reintroducing the temporality of surface dynamics, then allowing it to diverge from the viewer's temporality. The Eames Office—chiefly Judith Bronowski—spent many months on this problem, reading books and articles on relativity and considering various ways to depict time within the film. In reconstructing this history, I rely chiefly on notes, storyboards, letters, and scripts that were produced at the time and are preserved within the Eames Archive, a collection housed at the Library of Congress.

Ultimately, the filmmakers decided to incorporate several temporal indicators in their "instrument panel." Two analog clock faces represent "traveler's time" and "earth time," respectively. Each clock has ten numbers, each of which represents the passage of one second. One counter records the seconds that have elapsed for the "traveler," while a second counter gives the current speed of the vehicle as a percentage of the speed of light. This elaborate representational apparatus enables the Eameses to defer the temporal costs of a continuous zoom to domains other than that of the viewer, purportedly saving her from a space-time paradox.

Here's how it works: As the camera begins to zoom out, both clocks begin to tick off the seconds, perfectly synchronized. For every ten seconds of film time, the "traveler's time" counter increases by ten. The virtual traveler's timescale is thus aligned with the materially determined timescale of the filmic medium, and thus that of the viewer who watches it unfold. In each period of ten seconds that elapses for the viewer, 240 film frames are projected and ten seconds of time pass in the (virtual) milieu below. (The picnicking man is conveniently calmly asleep, so no movement is visible.) But by synchronizing and meticulously documenting the passage of time in these domains, the film has gained the capacity to document temporal divergence. As we approach the scale of the solar system, the "earth time" clock begins to run faster and faster, while the "traveler's time" clock maintains its steady pace (figure 10). "What seems a normal ten seconds to us would be a much longer period in relative earth time," explains the narrator. "Our clocks are getting out of synchronization." By the time the whole solar system is visible, the "earth time" clock is spinning wildly, and

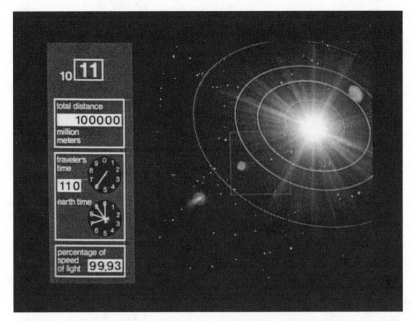

Figure 10 ▸ The dashboard's "earth time" clock can be seen spinning wildly. Frame from *Rough Sketch* by Charles and Ray Eames (1968).

the "percentage of speed of light" has increased to 99.99. The narrator informs us that "now a long time elapses on earth for each ten seconds of our traveling time. The clock is spinning round." Indeed, the second clock, as if its hands had spun off its face, suddenly disappears, replaced by a new indicator: "earth time per ten seconds." The digits of this new, digital clock keep track of how many hours, then years, then finally millions of years, are passing during each ten seconds of the film. Any indication of the absolute (total) time that has passed on earth has been removed in favor of tracking the change in velocity, or acceleration, of the "traveler." Because "earth time per ten seconds" continues to climb at ever faster rates, we know that this acceleration is continual and enormous. By the time we reach the largest scale depicted in the film, 100 million years are passing on earth for every ten seconds that pass for us (the traveler).

Why does the Eames Office go to such elaborate trouble to animate this complex instrument panel and track these diverging timescales? On at least one occasion during production a member of the team suggested either removing the instrument panel, reducing its size, or moving it to the less-prominent right side of the frame, away from the listing of powers and distance.[18] In the end, however, they kept in in place. I argue that it serves an absolutely vital function for the project, and represents the team's most ingenious cosmic-zoom innovation. It does more than merely diffuse the paradox of relativity toward which Boeke points. It is a primary mechanism in the work's refiguring of the scalar spectrum into a scalar continuum.

The problem faced by the Eameses was this: given the effects of relativity and the temporal constraints of the filmic medium (which unfolds at twenty-four frames per second), how could they produce anything like a continuous zoom that would encompass the entire known range of scales? Boeke hadn't faced this problem because the medium he chose, the codex, afforded an infinite medial unfolding and discrete jumps between scales. He responded to the relativistic paradox by eliminating time entirely from his medial apparatus, but readers confronting only one scale at a time could forgive this and imagine, through evocative textual description, the many interconnections between dynamics at different scales, and thereby produce a unified view of scale that did not require a visual or spatial continuum. On film, such a continuum could be a powerful visual motif of continuity, but only if *the rate of resolution change could be held constant.*

The Eameses' goal is explicit and announced at the beginning of the film's narration: every ten seconds of "traveler" (viewer) time will produce a change in *both* spatial and temporal scales by a factor of ten. The result will be a zoom that appears smooth and continuous: like a masterful camera assistant, it will alter our point of view in a smooth, steady, majestic alteration of our field of view. This cannot be achieved by moving our virtual vehicle at a steady rate. If, for instance, we were to travel straight upward at a rate of 100 meters per second (360 kilometers per hour), we would lift off of the ground with such a jolt that the human beings below us would be lost to sight within a couple of seconds. Worse, it would take us over a day of watching before the whole earth could be resolved! We would have the distinct advantage of witnessing events taking place in real time, but the zoom itself would not feel continuous: it would seem to start out very fast, slowing as we achieved higher altitudes, then blink by in an imperceptible period of time as we zoomed in to microscopic and smaller scales. Alternatively, scalar jumps could be made, as in *Cosmic View,* and temporalities could be established appropriate to the events taking place at those scales. With this technique scalar difference could be respected while distance, field of view, and time would all become discontinuous. The Eameses, like Szasz, were more interested in producing a continuum through time than in an exploration of discrete scales, and thus both chose a third route: hold the rate of resolution change constant and vary the relative timescales such that what we see at any time appears continuous, while that which lies below our threshold of resolution or outside our field of view is sped up or slowed down accordingly.

Szasz's continuous zoom takes place in a cartoon world that possesses no determinate relationship to real-world physics, perspective, or temporality. In such a world we can remove the effects of relativity. The Eameses, as their full title implied, wished to design a film about "the relative size of *things in the universe*" (emphasis added) and therefore worked with photographs rather than drawings to the extent possible, allowed time to run forward, and indicated through their instrument panel that the medial system of their film represented a *simulation* of the real world rather than an imagined alternative version. Moreover, Charles Eames's singular purpose in making such films, from *A Communications Primer* (1953) onward, was to teach mathematical and scientific concepts to the public. Judith Bronowski, who served as the chief researcher and designer of *Rough Sketch*, had been hired at least partially because her father, Jacob Bronowski, was a well-known mathematician and science popularizer admired by Charles Eames.[19] The Eameses' desire, then, was to constrain the film's representational system to known physical laws.

To hold the rate of resolution change constant within a modality of elevation change, a simulated camera must accelerate, changing its speed over time. To expand our field of view by a consistent factor of ten every ten seconds, we will need to increase our virtual speed by a factor of ten in the same interval. We can maintain this rate of acceleration for some time, but it doesn't take long before we are approaching a fundamental limitation: the speed of light. As Einstein elaborated, any object accelerated to near the speed of light will proportionally increase its mass such that the speed of light itself can never be reached. For the rate of zoom desired by the Eameses, this fundamental limit is approached by the time we resolve the scale of the solar system.

A very early draft of what appears to be the film's narration, scrawled in red pencil, begins with an evocation of relativity before launching into the description of the reference scale that opens the finished film: "The faster we travel the more slowly time seems to pass. / As we travel faster our mass increases to slow our approach to the speed of light."[20] By June 1968, when the first typewritten draft of the script appears, these two sentences have been excised. Nonetheless, the research materials of the production team are filled with relativistic calculations, and establishing a plausible model for the Eameses' desired constant zoom was clearly a high priority for Bronowski. She calculated, for instance, that a tenfold increase in velocity be-

78

tween scales was sufficient to stabilize their narrative zoom up to scale 8 (10^8). At scales 9 and 10, a reduction in acceleration due to increased mass would have to be offset by some other method. From 10^{11} onward, velocity would remain constant at 99.99 percent of the speed of light, and thus the zoom would lose its appearance of a constant rate if some other factor didn't compensate. What were the filmmakers to do?

Their solution was suggested by the special theory of relativity itself, which holds (to simplify) that the faster an observer travels, the more slowly time will seem to pass. In a popular example, an observer who leaves the earth and travels at a significant fraction of the speed of light will experience time normally, but she will age far less than her twin on Earth. Relative to this traveler, time on Earth will appear to have sped up drastically. The filmmakers count on this "time dilation" to secure a continued fixed zoom after acceleration has hit the velocity limit. A very early, handwritten draft of the screenplay included this explanation:

> If we continued this accelleration [*sic*] we would soon find ourselves exceeding the speed of light—and at that time our view would be frozen with one image because no new light waves could reach us to change our picture. So we are going to introduce a new concept. We are going to work with Einstein's Theory of Relativity. Relativity has shown us that the faster you travel, the more slowly time seems to pass. So we can say that we do not exceed the speed of light, for although the traveler experiences traveling a large distance in only 10 seconds, the effect is in fact a dillation [*sic*] of Earth time. The man lying in the park has experienced watching the traveller take much longer periods of time to cover the increasing distance.[21]

This curious formulation places the viewer in two positions: first, that of the traveler, who experiences only ten seconds passing between shifts in resolution, regardless of the distances involved. The vehicular motif and instrument panel superimposed on the viewing surface of this medial system have sutured us to the subjective position of this virtual traveler. She and we are said to be traveling so fast at this point that time has slowed down relative to Earth time, stabilizing our experience of the zoom as constant. From this perspective, the time it takes to traverse such vast distances is purchased through a speeding up of earth's time. This is the subjective position of the picnicking man, the second subject with whom we are asked to identify. He experiences the traveler (now othered from our point of view) as moving more and more slowly in her Eames spaceship. Relativity, as a "new concept" added to the simulation, enables the Eames Office to perform a temporal tradeoff from one domain to another, *from one subject position to another*. For the first and only time in the film, this segment of narration would have asked us to identify with the *reference* surface rather than

CHAPTER THREE

the medial surface. The effect, however, is to dramatize the medial system's temporal bifurcation as the diverging experiences of two human-scale observers. While this may function well in the context of relativity thought experiments, here its effect is ironic: in order to secure the experience of a constant zoom for one observer, the medial system has established a relativistic dilation by introducing a second "observer" to whom such an experience is expressly forbidden.

Picnicker or traveler? *Rough Sketch*'s visual semiotics produce competing claims on the viewer's subjective identification, or suture.[22] The opening scene of the film centers the perspective on the white, male picnicker (rather than his companion, the marginal woman). Indeed, this character remains literally centered throughout the film, and it is his body we enter during the "inward journey." On the other hand, the first-person perspective suggested by the instrument panel and narration lays equal claim to the viewer's subjective identification. Picnic-time or travel-time? The viewer, faced with two bifurcating temporalities, can do naught but jump between the competing subject positions, thereby disrupting her own sutured continuity with the subject position that is the beneficiary of this dilation. To experience a continuity of resolution change, the viewer must both remain sutured to the position of the traveler *and* temporally diverge from the surface she is viewing as the traveler. It is this paradox, a discursive paradox, that the filmmakers grappled with in the first half of 1968. As with the previous Einstein reference, by June this narration had been excised from the script, weakening the picnicker's claim on the viewer's subjectivity and further cementing the interlocking of the viewer's timescale with that of the traveler. The picnicker's timescale, represented only by the second of the relativity clocks, recedes into abstraction without the original narration.

The basic strategy of relying on a mechanism of relativistic time dilation to transfer the temporal costs of a zooming continuum from the viewer's timescale to the observed surface's timescale, along with its attendant temporal-subjective paradox, nonetheless remained. Bronowski discovered in the course of her research a 1966 book by geophysicist Sam Elton titled *A New Model of the Solar System*, which may have provided support for such a semiotic strategy. One of three quotes she copied out of the book reads, "The idea of compensating for 'large size' by using an 'enlarged time scale' may be an essential ingredient in formulating a recipe that will make a unified universe."[23] Elton's concept is both elegant in its formulation and remarkable in its diction. Immense physical scales can be speculatively tamed by enlarging our timescales. Elton's goal in *A New Model* is to advance a historical model of the solar system, positing that every cosmic body is in the process of transforming into other bodies: moons into planets, planets into gas giants, gas giants into stars, solar systems into galaxies, and so on. This process is cyclical: each local entity has a history, but the system as a whole is in a steady state. "Within our observable sphere of the universe we find that all the units evolve while the general distribution of galaxies remains the same."[24] Elton contends that his theory seems counterintuitive for two reasons: cosmologists have narrowly clung to the view that all of the matter in the universe was created in an instant of time and isn't changing, and our temporal point of view is so narrow that we cannot see the larger changes that are occurring to all bodies, regardless of scale.

Elton's solution is to ask his readers to imagine the universe as a motion picture compressing eons of time into a half hour. If we could only imagine (or simulate) this film, we could begin to conceive of a dynamic universe in which bodies grow from youth to old age, interact, and finally die. Elton's emphasis is on the capacity of the human mind to conquer other scales by trading space for time. It is no surprise that Einsteinian thought would serve as the conceptual tool *par excellence* for this project, even if the introduction of this "new concept" would ultimately only be allowed, for the Eames Office, to take the form of diverging clocks in a virtual instrument panel. Elton's vision, shared by the Eameses, is one of a unified universe (seemingly a redundancy of terms). Such a unification is not discovered but "made," constructed out of the potentials unleashed by spatial and temporal difference. The spoils go to the designers of medial systems that can dynamically link spatial and temporal multiplicities and thereby enable transmutations between and among them.

Elton's "unified universe" milks Einsteinian relativity for everything it's worth, trading time for space in a kind of microlending scheme aimed squarely at the human-scale reader. The unfathomably large—planets, galaxies—may appear to belong to a different order of things than those that populate our ordinary milieus: a monstrous, unchanging, nonhuman order. Actually, however, these things are a lot like human things. To gain this perspective, to triumph over and through the cosmic sublime, one need only enlist the help of cinema's bifurcated timescale: film the planets for a very long time, then speed that film up as you shrink it down, and by the time it reaches human scale it is ready for a matinée showing at the local cinema.

Elton's point is media-savvy: scale can be mediated. His cosmic film camera may be virtual, but the result of screening this mental film is precisely to produce a virtual perspective, an idea of the universe. This is a universe that can be fully visualized through human mediation, provided that time and space can be freely exchanged to transpose scale. It is a metastable universe, in which every scalar milieu changes cyclically, but in which no scale propagates linear change to any other scale: the system as a whole never changes. This metastability is apparently authorized by yet another property of the universe: its various scales all change according to the same dynamics, follow the same cycles. Planets, solar systems, and galaxies all turn out to follow the life cycle of the human in this remarkable formulation, rhetorically hearkening back to the alchemical theories of microcosms propagated in late medieval Europe.[25] Little in Elton's assertion diverges from the theories of the microcosm advanced by the alchemist Paracelsus in the sixteenth century: "External stars affect the man, and the internal stars in man affect outward things, in fact and in operation, the one on the other. . . . Thus are the double stars related one to the other. Man can affect heaven no less than heaven affects man."[26]

Elton's concept of a cosmic life cycle that mirrors that of the human body, rhetorically supported by an appeal to cinema as a mediating timespace transmutation, is a textbook example of what I refer to in chapter 1 as "scalar collapse," the conflation of two or more scales through a media technology that speculatively rescales them into the same visual or conceptual plane, replacing the dynamics of one with the dynamics of the other. Here the solar system comes to behave like the human body. But where Elton merely invokes cinema as a conceptual device to effect the rescaling—trading space for time—Bronowski literalizes the rhetorical move, extracting Elton's quote from his extended thought experiment and operationalizing it in the making of *Rough Sketch*. This changes the valence of unification. For Elton, this rhetoric reveals a "unified universe," that is, a universe unified by anthropocentric subjectivity in its scale-mediating fullness. The collapse of scale authorizes such a unification, but the human mind, propped up by its tranmutational media, effects it. The universe, its scales rendered equivalent through scalar collapse, is woven together by human consciousness into a single continuous surface. This is an analog universe, unburdened of discrete transitions, uncut by scale.

Rough Sketch makes no direct appeal or argument for an analog universe but uses the filmic medium to produce a powerful visual evocation of exactly this. Smoothly zooming from scale to scale in real time (relative to the traveler), while sweeping under the seats the radically accelerating time dilation animating this overdriven cosmos, sutures all scales together into a continuous contour traced by and for the human observer. The film's analog clocks proudly, if honestly, reveal the true medial engine powering our frictionless journey. The Eames team considered this a sufficient solution to the problem of acceleration threatening to undermine the seamless-ness of the viewer's travels. But if we take the invitation of the film's ill-fated draft narration and examine this solution from the point of view of our sleeping picnicker, it appears in a different light: namely, as the progressive acceleration of the speed of the system (the simulation) during the outward journey of the traveler, and the progressive deceleration of the simulation during her inward journey.

Ironically, the traveler is progressively disarticulated from the system she is meant to observe. If in this regard she plays the role of the paradig-matic Western subject, transcendent and apart from the world she inhabits and studies, we must not forget the displaced role played by our picnicker. The traveler's experience of a smooth zoom to the outer edges of the uni-verse and back again have cost *him* over a billion years. A truly unified universe is purchased at the cost of irreconcilable temporal scale bifurca-tion. We may well wonder how it is possible, then, that at the end of this extended vacation our two long-separated friends could meet again, as if a mere five minutes had elapsed! What scales of dreams our picnic-bound somnambulist must have gleaned, what dreams of scales.

TOYING WITH IDEAS: THE SCALAR ANALOG

In 1970 film critic Paul Schrader—long before he became a celebrated screenwriter and director—published a most unusual essay in *Film Quar-terly*. Its title was "The Films of Charles Eames." On the surface Eames was an odd subject for a retrospective in a top film journal: an architect by training, he was world-famous for the revolutionary lines of furniture he designed and marketed with his wife and partner, Ray. Their house, the Eames House, served then—and now—as an icon of postwar architectural modernism. But their films?[27] Most were a few minutes long, and only a couple had crossed the fifteen-minute threshold. They were mere compo-nents of exhibitions designed by the Eames Office, or animated educational shorts commissioned by IBM to help make computers and mathematics seem more friendly to ordinary people. Some were simply aesthetic exer-cises documenting the washing of an asphalt schoolyard or the traditional folk art objects produced in Mexico during Día de los Muertos.[28] They were intricately, even masterfully, crafted, but curators would be as unlikely to show them in a gallery as proprietors would be to show them in cinemas. One could find them projected in classrooms or pavilions, but these are decidedly not the regular haunts of film critics. Yet here is Paul Schrader, brazenly treating this body of work as narrative cinema. I wish to linger on his essay, as it provides us with an entry point into the expanded notion of the scalar analog that I am developing here, a key to the analog nature of the Eames films.

Schrader begins by asserting that "Eames' films do not function independently, but like branches."[29] That is, they are best treated as a collective, rhizomatic body with multiplicitous expressions. Nonetheless, he writes, they fall into two great classes: "Toy Films" and "Idea Films." The former exhibit "object-integrity," that is, they respect objects for their design: their pure shape, color, and movement. The exemplary work here is *Tops*, a seven-minute film from 1969 that consists only of a collection of antique tops from around the world, shown in a three-stage montage—first being wound and released, then spinning, then collapsing at that singular moment when their angular velocity can no longer keep them upright. The second category, the idea film, is exemplified by a series of two-minute "mathematics peep shows" created for IBM in 1961. For example, 2^n: A Story of the Power of Numbers demonstrates, though an animated narrative, the non-intuitive implications of calculating with exponents. In an Indian palace, a grand vizier introduces his king to the game of chess. When the delighted king offers him a reward of his choice, the vizier's request sounds humble enough: a single grain of wheat on the top left square of the chessboard, followed by twice that number on the next square, then twice as many again on the next, and so on, until every square on the board is filled. The king laughs and orders it done. But by the forty-second square, much to his chagrin, the entire palace is filled with wheat! He realizes that the court gamer has duped him with another game entirely. Had the process been completed, it turns out, wheat would have "covered all of India to a depth of over fifty feet," an eventuality illustrated in the film's animation. That quantity of wheat grains, remarks the narrator (Charles Eames), would be enough to stretch "from the earth, beyond the sun, past the orbits of the planets, far out across the galaxy to the star Alpha Centauri, four light years away. They would then stretch back to the earth, back to Alpha Centauri, and back to the earth again" (figure 11).

This tongue-in-cheek parable warns us against relying on our intuitive sense of measurement when powers are involved. Multiplying a number by itself produces very large quantities in a relatively small number of iterations. As the Eameses saw clearly, all that remains is to couple an iterative-powers mechanism to a constant time-based medial form and you have a machine for animating radical transformations in a given calculable domain over short periods of time. In an actualized version of this scenario, every successive deposition of wheat would take longer by more than a factor

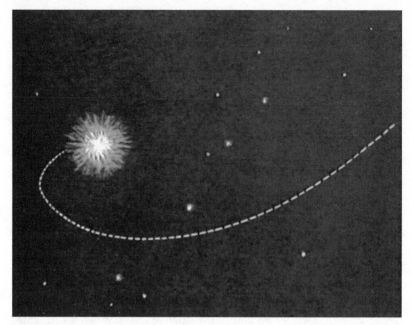

Figure 11 ▸ Frame from 2^n by Ray and Charles Eames (1961).

of 2^n (as new methods of procurement and transportation would have to be devised and constructed for each successive quantity), and thus the king would be saved, ultimately, by time. He might go bankrupt, but he would not find himself swimming in wheat during his lifetime. In 2^n, however, this is exactly the situation depicted, a comic effect that is afforded by the filmic timescale, here aligned with the iteration of the powers calculation rather than the unfolding of its consequences. Like the later *Rough Sketch*, 2^n achieves its visual effect only by relativizing time. This is not, in fact, the only "cheat" in the film. The mathematical consultant employed on the film, Raymond Redheffer, pointed out during its production that there was an error in the filmmakers' calculations, and the final depth of the grain engulfing India would only be six inches.[30] Charles Eames chose not to heed Redheffer's advice, sacrificing scientific veracity in a rehearsal of the enabling cheats utilized in *Rough Sketch* seven years later.

The scalar politics of 2^n are clear: radical scale shifts incur no temporal cost. The physical dynamics of any nonhuman scale (in this case a subcontinent-size pile of grain, or an interstellar grain elevator) are subordinated to the abstract theoretical calculations of the film's protagonist. This scalar collapse, mirroring Archimedes's argument that mathematics can outscale any physical entity (see chapter 1) is entirely expected in a film created for an exhibit on mathematics, produced with sponsorship by IBM—a corporation committed to selling exactly such computational power fantasies. 2^n, then, serves not only as a warning to mathematically uninclined monarchs but as a sort of manifesto for the medial potentials of "power thinking." I borrow this latter term from Hans Christian Von Baeyer, who argues that we are entering an age in which measurement has achieved such radically disparate scales that our intuitive models will no longer be adequate for understanding the world. In order to make large numbers small again, we must begin to think in powers, or in terms of logarithmic scales. "Nothing less than a transformation of quantitative thinking will be required, and the logarithm is just the tool for the job. Power thinking, the stock in trade of cosmologists and elementary particle physicists, will have to become a common habit of mind, lest we all drown in a sea of digits."[31] Drown, we might add, like the Eameses' cartoon king, engulfed by his outdated paradigm.

Schrader suggests that in the Eameses' filmic oeuvre, the primary difference between the toy film and the idea film is one of information density. In the precomputer world of craftsmanship, the object-integrity of the toy is a question of its material surface properties. In the computer age, object-integrity necessarily involves a more complex aesthetic: "information overload." This aesthetic arises from an overabundance of information that cannot be whittled down and painted into its essential characteristics in the form of a toy. Instead, the saturation of information that the computer has wrought must be processed and navigated into its own sort of quintessential object: the idea. Thus, in an Eames film,

> The viewer must rapidly sort out and prune the superabundant data if he is to follow the swift progression of thought. This process of elimination continues until the viewer has pruned away everything but the disembodied Idea. By giving the viewer more information than he can assimilate, information-overload short-circuits the normal conduits of inductive reasoning. The classic movie staple is the chase, and Eames's films present a new kind of chase, *a chase through a set of information in search of an Idea.*[32]

In this formulation, each idea film presents us with multiple dimensions of supposedly unassimilable data, but then marches us through that data at the timescale of the film's unfolding, forcing us through a subtractive process to grasp a quasi-Platonic idea that holds it all together, that provides a basic model that can be unfolded into the pattern that governs the film's structure. The idea is the key that unlocks not only the film's overarching logic but also the contours of a path to follow through an information-saturated milieu. As such, the idea is a model for navigating the world, regardless of the singular features one's environment may present. It is a meta-environment, an aesthetic that can be applied to the world at large. Not surprisingly, then, Schrader reads *Rough Sketch* as the idea film *par excellence*. Here the bewildering complexity of a literal universe of information is reduced to the smooth field of view continuum I discussed in the previous section. The idea, according to Schrader, is "the powers-of-ten-extended-through-space-and-time."[33] That extension is, I have argued, an analog contour articulated to the human subject as observer and mediator. Just as low-level processing is hidden below a graphical user interface, digital computation links radically disparate scales through logarithmic series, trading time for space, while the subjective foreshortening that this enables traces an analog path through a universe teeming with overwhelming detail. The filmic idea, which is both procedure (algorithm) and aesthetic (contour), manages to underwrite the experiential dimension of the latter with the medial logic of the former.

As we have seen, acceleration and temporal bifurcation align the smooth expansion and contraction of field of view with the linear, regulated temporality of the film's projection speed and the time of the viewer. All of this is synchronized and made mathematically possible through iterative applications of power thinking. As students first discover in their trigonometry classes, powers of ten offer particularly effective shortcuts to perform calculation on large numbers: to multiply or divide a number by ten, all one need do is add or subtract a zero or, in logarithmic notation, add or subtract one power from the base number. As the base number here is 1 (meter), the Eameses have justified removing it entirely, reducing the "distance" measurement in their control panel to only the power itself (10^{16}, 10^{17}, etc.). This shift is emphasized on the dashboard's instrument panel through a dedicated display of powers that count steadily up and then, during the inward journey, down to negative powers.

This represents an innovation. Boeke's *Cosmic View*, primarily concerned with the material relationship between the page and the reference surface, displays the scalar ratio in linear numerical form—including all zeros. Szasz's *Cosmic Zoom* omits all numbers. The Eameses, however, prominently display powers in both *Rough Sketch* and the later *Powers of Ten*.[34] The narrator reinforces these "pure powers" by repeating them often, even formulating them into a countdown during the inward journey. The combined effect is almost shocking: the entire world becomes decomposable in terms of regular scales that can be easily navigated with a simple mathematic manipulation. Better yet, a logarithmic acceleration coupled with a linear medial temporality allows the scales to be navigated *automatically* as the output of a clockwork medial machine: just wind it up, sit back, and relax. The powers of ten will provide the fuel for our toy spaceship. While Schrader is certainly right that this is an idea perfectly formulated in filmic terms as a mechanism to reduce complexity, I would suggest that *Rough Sketch* and *Powers of Ten*, in their clever mechanisms and bold presentations of crafted surfaces, owe just as much to the Eameses' toy film genealogy.

As we have seen, Boeke's *Cosmic View*, despite its relatively crude drawings, renders each scale in great physical detail, largely through the interplay of image and text. Scales are milieus with inherent dynamics and characteristic objects, interconnecting in various ways. No fewer than thirty-seven of *Cosmic View*'s scales—more than 90 percent—contain this sort of dynamic detail. Most of this ecological detail is lost in the Eameses' cosmic-zoom films (there are exceptions, as we shall see in the next chapter). Yet these are just the sort of details that feature prominently in their toy films. In *Tops* the material construction of the tops is shown, without narration, to correlate to certain physical states and dynamic patterns of motion and color. Part 1 of the film serves as a catalog of mechanisms for transferring radial motion to circular objects. Part 2 is a dance of colored patterns. Part 3 documents the correlation of shape and size to possible collisions between surfaces.

According to Philip and Phylis Morrison, consultants on *Rough Sketch* and collaborators on *Powers of Ten*, *Tops* "is always well received by physicists and astronomers, who find in it examples of the same spin that is everywhere, in sea and air, in planets, comets, stars, galaxies—in every proton." In other words, there is something trans-scalar about this film, evoked by its miniature tops in their elegant but raw materiality. "Without symbols it transmits a depth of instruction in real science that is hard to match, a charming view of one striking portion of physical reality, transmitted with unspoken but gripping internal drama made plain by the sharp and steady view."[35] The Morrisons are not the only ones to implicitly link *Tops* with the Eameses' cosmic-zoom films. In 1970, when Charles Eames became the Charles Eliot Norton Professor of Poetry at Harvard University, a position that entailed delivering six lectures that academic year, he built his first lecture around a joint screening of *Tops* and *Rough Sketch*. The topic of the lecture was "making choices" and the importance of working out problems as part of the pleasures of everyday life.[36] Both films tackle the design problem of spanning scales to examine material processes outside of the ordinary human spatiotemporal framework—that is, of capturing events from nonhuman perspectives.

Another remarkable Eames film, *Toccata for Toy Trains*, reconstructs an

Figure 12 ▸ A non-embarrassed toy train, from *Toccata for Toy Trains* by Charles and Ray Eames (1957).

entire miniature world of hand-crafted toys from different eras. Through various physical mechanisms, the Eameses animate all of the toy trains, cars, traffic lights, and pedestrians that make up this alien but intricate milieu (figure 12). The film, released in 1957, took over a year to make, the miniature set constructed within the studio at the Eames House, before Ray and Charles moved their filmmaking operations into the Eames Office.[37]

Charles Eames himself provides the film's narration, expressing his philosophy of toys. Categorically, he explains, they are different from scale models. Scale models are simply miniature versions of their macro counterparts. Toys, on the other hand, "have a direct and unembarrassed manner that give us . . . a pleasure different from the admiration we may feel for the perfect little copy of the real thing."[38] Toys proudly display their difference from the macro-scale objects whose form they share. Thus a wheel may be constructed out of a button instead of rubber, a car out of wood and plastic that would be inappropriate on our highways. This, suggests Eames, is the secret to the pleasure of toys. Scale models may be wondrous in their construction, but they fail to convey the aesthetic pleasure of the toy, which trades on scalar difference to combine the aesthetics of the macro scale (its forms, its colors, its rough-hewn materials) with the size of the miniature.

This material-aesthetic clash produces what we might call the pleasure of trans-scalar play. We would expect a perfect scale-model train to move under its own power, but as any two-year-old knows instinctively, that relatively intellectual pleasure cannot compare to the kinetic pleasure of pushing these slightly monstrous trans-scalar hybrids along the floor. This is what reminds us that they are not autonomous, that they rely upon the dynamics of a larger scale to give them life. As such, toys inherit the time-scales of their material constituents. They represent multitemporal assemblages, concentrated into a particular set of scalar coordinates and unified through play.

Rough Sketch and Powers of Ten represent a synthesis between ideas and toys, with power thinking acting as the fulcrum upon which both turn. Iterative logarithmic manipulation of all of the factors we've discussed, intermeshed like clockwork into a cinematic treatment of scale—this is the prelude to the greater idea, which incorporates the scalar aesthetics of toys. The idea of the trans-scalar zoom, of a movable point of view that can encompass and contain all scales as a clockwork mechanism: this is the real achievement of the Eameses' cosmic-zoom films. This is both an idea and a toy, an idea of a toy manifested as the toy of an idea. It is a model of the earth, the galaxy, the atom. Not a miniature version, which would contain the dynamic complexity of its larger counterpart as a microcosm, but a toy version that leaves the ecological detail of the original unresolved and bears instead the mark of deliberate design—that is designed, in fact, for play.

The referent of this dynamic toy (a toy only becomes a toy when one plays with it) is not any given scalar milieu or all of them together, but a subject: the human being. Every scale that emerges through the watching of Rough Sketch is a reminder of the mathematical manipulation (adding or subtracting one power), visual perception (the human gaze), and design process that together stabilize and set in motion the trans-scalar zoom. The human is the privileged scale: it is more than the reference scale that opens each film, it is the scale of design, the purpose of design. The Eameses' best-known chair was a breakthrough because it expressed the industrial material of plywood in a form that fit the human body. As Charles Eames notes, design is "a plan for arranging elements for a particular purpose" that "addresses itself to the need."[39] Justus Nieland has argued that the Eameses' design ethos was explicitly scalar, "attending to the senses' remaking in man-made environments and technical networks that now exceed humans" while also seeking to "reorient, and rescale the human domains of perception, attention, and care" in response to global resource demands and newfound awareness of large-scale systems."[40]

And yet, while Eamesian design, especially in the cosmic-zoom films, may share Boeke's twin goals of inculcating "a sense of scale" and efficiently managing resources on a global scale, the Eameses' toy films celebrate design deployed in the opposite scalar vector. A top is designed to accept human-generated kinetic energy, transferring it to materials on a smaller scale. *Tops* documents and seeks to convey the pleasure of this scalar transfer, during which the enormous, human-scale hand barely dips into the frame.[41] The magic moment, repeated numerous times, when two surfaces make contact, is a return to the human scale, a transfer of energy back to the sustaining surface of the tabletop or floor. A moment of scalar friction resolved only within the framework of human play. *Rough Sketch* follows this basic formula, producing continuous accessible surfaces that always remain in direct speculative contact with the extensible human. Scales are explored not as nested milieus that recede farther and farther from the reach of human design with each jump, but as a single contiguous surface that never varies in distance. The human hand is, as in *Tops*, omnipresent, always within range of touching. This grand toy, the trans-scalar zoom, is species-specific.

The Eameses' disdain for the miniature and reverence for the similarly scaled toy is telling. The toy, in all its rough-hewn splendor, is consummately analog. It bears the marks of its own trans-scalar construction: there is a mismatch between the scale of its reference object and the scale of its constituent materials. Those materials, or rather their signifying dimension, always bear a scalar trace back to the everyday life-scale of the human, whence the corks, buttons, and rough-hewn wood blocks originate. The toy also signifies, according to Susan Stewart, "both experience of interiority and the process by which that interior is constructed."[42] The toy, as signifying surface, relativizes its constituent scales to the temporal and spatial scales of the human subject. In other words, the toy is a kind of media that both resignifies other scales as interior landscapes for the human and makes visible its own mediating qualities. For the Eameses, this process is a mechanism by which design space, or instrumentalized creative development, extends its domain along the scalar spectrum. The material construction of the toy is foregrounded, while ecological detail—complexity—is eliminated.

You don't need a magnifying glass to resolve a toy's material affordances.[43] Despite the charm of its exaggerated features, the toy represents human domination of the environment rather than its reflection, as in the miniature replica. The idea, on the other hand, is a speculative conjoining of complexity-pruning mechanisms (algorithm) with aesthetics. In practice, the chessboard buried under a heap of wheat stands as an adequate icon of the idea. And here I propose to modify Paul Schrader's filmic taxonomy: these two types of Eames films, rather than constituting separate or even incompatible trajectories, are actually complementary. Together, the digital-buried-under-the-analog idea and the design-substituted toy produce a bifurcated temporality and a toy aesthetic with which to navigate the cosmos, come what scale may.

And yet, *Rough Sketch* cheats as a toy. It fails the Eameses' own test: it is not unashamed of its toy-nature. This is where toys can become dangerous.[44] The problem that *Cosmic View* sets out to solve is one of scalar alterity: how to understand the cosmos when it is composed of scales that elude and alienate us? The Eames Office addressed this question again and again from the late 1950s through the 1960s. In both *Glimpses of the U.S.A.*, a 1959 project in Moscow, and *Think*, a multiscreen exhibition at the 1964 World's Fair's IBM Pavilion, they used fragmented images projected on the inside of a dome to produce a media environment that brought the scales of the larger world into direct contact with the human viewer. Fred Turner identifies both installations as exemplars of what he calls the "democratic surround," a large-scale, postwar media project using designed immersive environments to cultivate an American democratic subject. Turner's description of *Think* applies equally well to *Cosmic View*, released four years later:

> The Eameses' projections at the IBM Pavilion literally lifted people out of their ordinary lives and placed them within a network of images . . . the show's overwhelming multi-screen, multi-sourced-sound array asked viewers to do the perceptual work prescribed by Herbert Bayer of selecting and integrating sound and images into their own individual psyches. In that sense, it asked viewers to become information processors.[45]

Think's solution to the problem of incorporating alien scales into individual subjectivity was to surround the individual with a panoply of images from elsewhere that were, despite their scalar difference, to be integrated by virtue of their spatial and temporal continuity and the directed attention of the human nervous system, along with its natural ability to assimilate and connect.[46] *Cosmic View*'s solution to the same problem is schematically similar: unite images from different scales through coterminous space and time. This solution is a digital-analog hybrid: incorporate power thinking to compute ligaments between scales, a sort of boardwalk through the scalar spectrum, and render it as an analog trail. The smoothing over of difference that this entails requires a burying of even its own calculating magic. On the animation stand, the process occludes its own dynamics of

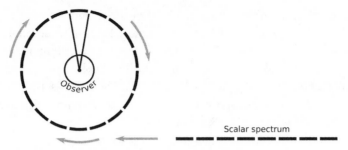

Figure 13 ▸ Inside the zoetrope. The scalar spectrum as sutured, equidistant, analog wheel.

resolution, burying the jumps between distinct scales, distinct surfaces, under a series of cross-fades. This is a secretive, ashamed toy that hides its own seams—even as it hides the scalar seams of the universe it traverses.

The vexed calculations and discussions of relativity that animated the Eames Office's construction of *Rough Sketch* remind us that spatial scale and temporal scale cannot be disarticulated. Not only is each inextricably bound up with the other, but mediating apparatuses, which transpose scalar relationships into surface relationships and introduce both a time and space of the spectator, can manipulate spatial and temporal scale by trading them off against each other. We might call this *medial scalar relativity*. If the Eameses' surrounds spatially unite images and temporalities from elsewhere, their cosmic zooms accomplish something similar by trading time for space, bringing images (scales) from elsewhere into a coterminous viewing position. It may seem that the spatial surround has been replaced by a single, focused image source, but the surround is still present, just not simultaneously. The scalar spectrum encircles the viewing subject even if only a narrow portion of it is visible on-screen at a given instant: a digital enframing of an analog trace of scalar difference.

The cosmic zoom, in its classic cinematic form, as established by the Eameses, composes a model of the universe with its own temporal and spatial unfolding. *Rough Sketch* represents not merely an analog model but a model *of an analog universe*, wherein discontinuities within the scalar spectrum have been sutured by the gaze of the human into a continuous band encircling the subject. Like all models, it trains its viewers in how to approach temporal and spatial configurations outside of their own immediate scalar milieus. That is, in *Rough Sketch*'s analog universe, its scalar coordinates are all equidistant from the human subject they stabilize. The viewer appears to be moving through continuous space, but actually, constructed space is wrapping around the viewer, always at the same distance, aesthetically and epistemologically (figure 13). This is the zoetrope model of scale, a spinning wheel that it is fittingly both medial toy (a window "into" all other scales) and an idea—of the human. The human subject, as knowledge consumer and producer, stands upon a fixed platform that serves as a model of knowledge itself. Accordingly, it is to the disciplinary mediation of scale that I turn in the next chapter.

shaping scale

Powers of Ten and the Politics of Trans-Scalar Constellation

ONE OR MANY SURFACES?

Powers of Ten, the most famous film produced by the Eames Office, seems on its surface to be merely a polished and updated version of *Rough Sketch*. Released in 1977, it appears to follow the same structure and drama of resolution: two humans, a marginally positioned woman and a central man, are having a picnic in a park. They fall asleep and an overhead camera begins zooming out. We see the surrounding city, harbor, region, planet, solar system, galaxy, and galactic cluster in sequence, each scale along the way framed by a superimposed square delineating its boundaries and a "dashboard" indicating the length in meters, in powers of ten, of that bounding box's sides. A narrator (now male instead of female) characterizes the scales we encounter, up to the entire universe and down to the quarks inside a carbon atom. The location has changed from the Miami golf course of the earlier film to a Chicago park on the shore of Lake Michigan, but the picnic scene itself seems unchanged. Indeed, Ray Eames took great care to replicate the picnic, including its props, as accurately as possible.[1] The actor playing the man is even the same—Paul Bruhwiler, a Swiss designer who had been a staff member of the Eames Office when the first film was made, happened to be visiting Los Angeles in 1977, and was thus able to reprise his role.[2] Yet for all of the similarities between the two films, at their core they present radically different models of the universe. The nature of that difference, and of trans-scalar constellation—the particular arrangement and characterization of the scalar spectrum that every cosmic zoom per-forms—is the subject of this chapter. As we shall see, the choices made by the designers of cosmic zooms can radically alter our perceptions of—and thus our engagement with—the world around us.

By the time *Rough Sketch* was completed, its limitations were already manifest. The production team had left unresolved the temporal problems discussed in the preceding chapter. At the film's first public screening, for the Commission on College Physics, a group of electron microscopists found the inward journey to be unconvincing.[3] In retrospect, it seemed that the Eames Office had paid far more attention to the outward journey, scientifically and technically. The film proved successful as a proposal, however, and longtime corporate client IBM agreed to fund a more elabo-rate version.[4] Meanwhile, word about the film spread within the scientific community, and the Eames Office received a handful of purchase requests. They obliged any requests they received, selling a number of 16-millimeter prints to universities and scientific societies.[5] As they began work on *Powers of Ten* in the early 1970s, however, they halted sales of *Rough Sketch* and asked inquirers to wait until the new film was available.

Having obtained IBM's sponsorship, the Eameses quickly expanded their list of consultants and hired new research and filmmaking teams, headed by a promising young animator named Alex Funke.[6] This new working group eventually consisted of nine consultants, ten image (art-work) producers, three filmmakers, longtime collaborator and famed film composer Elmer Bernstein, and the Eameses themselves. Their task: to produce what was now referred to as *Powers of Ten*, a substantially updated and more realistic version of *Rough Sketch*. One of the team's first actions was an evaluation of the earlier film. The members noted many of the prob-lems discussed in the previous chapter, as well as critiquing the artwork. Of the chromosomes shown at 10^{-7}, one researcher writes, "No one knows what genes look like. They are certainly not these doughnut shapes." The scale of the atom (10^{-10}) comes under similar scrutiny for misrepresenting the outer electron shell as a surface like that of a ball of yarn rather than as open space. They also note that the photo of the earth isn't correct: "This photo doesn't appear to be over Florida. Either we get a photo with more clouds or an accurate one."[7] Finally, they note "conceptual problems" raised

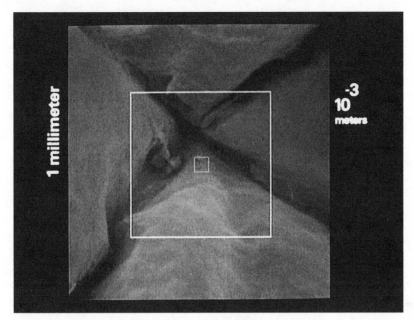

Figure 14 ▸ Approaching the surface of the skin. Frame from *Powers of Ten* by Charles and Ray Eames (1977).

by the apparent movement *through* the skin at 10^{-3}. The film supposedly functions on the assumption that the viewer's field of view is changed by moving closer or farther away from the central object. If that is the case, this object should be located on a surface accessible, in principle, from every scale, yet our virtual vehicle somehow reaches a white blood cell *within* rather than on the surface of the picnicker's hand. This introduces a spatial—and thus conceptual—discontinuity at the surface of the skin. The new production team suggested that the target object be moved to an external surface: either a skin cell or a blood vessel on the surface of the eye.

This concern about the introduction of more than one surface to navigate is significant. As I argued in chapter 2, it is *surfaces* that are resolved. Because scale, according to my working definition, is a medial negotiation of difference between surfaces, a complication is introduced if the primary surface is bifurcated or multiplied. This is because a scale becomes legible as the relationship between the reference surface and the medial surface is stabilized. That relationship requires that the two surfaces themselves remain constant. Just as a sudden change of viewing surface (if someone were to yank away the projection screen and replace it with a computer monitor, for instance) would necessitate a new scalar negotiation, so would the jump from one reference surface to another (figure 14). The problem, as the production team notes, is conceptual: the monadic surface alone undergirds the spatial continuum of the cosmic zoom. If that surface changes during the course of the zoom, what authorizes us to assume that we have not jumped scales? This was not a problem for *Cosmic View*, which Boeke explicitly presents as a series of "jumps." But the overriding preoccupation for the Eameses was scalar *continuity*, and thus they took great pains to construct their cosmic zoom as undergirded by a *single* scalar negotiation. Any element that could threaten the harmony of this conceptual zoom through all known scales was rigorously scrutinized and debated within the production team.

In this case, the problem never really dissipated. The Eameses wanted to keep the hand—perhaps as a symbol of human technics—and so a proviso was inserted in the script: at the beginning of the inward journey, as the narrator explained that in each ten seconds we will now narrow our field of view by a factor of ten, she would add, "We agree to assume that opaque substances, like the skin, will become transparent as needed, so that we can see into them without destroying anything."[8] This partially solves the problem by magically altering properties of the reference surface's features to enable strategic scopic access while maintaining the unity of that surface. A robust understanding of the surface must take topology into account.[9] This solution, then, introduces a new transparency rule: surfaces can become transparent in lieu of us traveling through them. The tradeoff is that our travel is suspended: our disembodied point of view continues onward, penetrating surfaces through an optical metaphor where it can no longer make use of a positional metaphor.

The circumstances of the "transparency rule," however, reveal a fundamental difficulty of the cosmic zoom: while arcs of resolution and mechanized temporality can be skillfully conjoined to produce a single contiguous, equidistant surface of epistemic access for the viewer, the *natural media* of the actual world—the reference surface—contain discontinuities that disrupt the visual modality of the continuous zoom. This is because, to paraphrase this book's first chapter, scale is mediated, but media is also scaled. The problem highlighted here isn't a technical problem of visualization but a conceptual problem that confronts the properties of a particular scale, properties that elude representation as a scalar spectrum. Skin, a substance made of fatty tissue encasing a muscular-skeletal structure, is scale-singular: it can be resolved only at certain scales, and such resolution reveals a thin, pliable membrane that is porous yet not transparent. It is a barrier, and a particularly significant one that maintains the integrity of both organs and organisms, hides the inner state of the poor dissembling creatures it adorns, and instead reflects back the codes of its social milieu (race, age, beauty, etc.).

To say that skin is scale-singular is to say that, while it is caught up in all sorts of trans-scalar dynamics, there is one narrow band of the scalar spectrum upon which it can be resolved qua skin (instead of qua organism, qua hide, qua racial signifier). It becomes skin again when we can make out its pores, its unique patina of stretch marks, the microcontours of veins and cartilage to which it so closely conforms. Its scalar function: to enable certain effects at certain scales and to act as a prophylactic against others. Water molecules may pass given long enough contact, but unicellular organisms cannot, most viruses cannot, and other skin cannot—this marks the boundaries of the organism in its encounters with others. Skin, then, participates in enormously complex multiscalar economies and ecologies, but has a precise home on the scalar spectrum. Certain events occur only at its scale (the mosquito bite, for example, utilized in both *Cosmic View* and Eva Szasz's *Cosmic Zoom*).

Skin is special. This is partly because it makes ecologies at different scales possible, and partly because we, as subjects playing language games, attach particular meanings to it. But what if we were ask about its generic scale, 10^{-2}? Is that scale special, just because that's where we find skin? Attempt to resolve any organism at 10^{-2} meters and you will run into the problem the Eameses did: skin will get in the way. Resolve a rock and you will find a pattern of microfissures. Resolve a plant and you will find the characteristically fibrous patterns of cellulose. Wherever you look, 10^{-2} meters is a scale at which events occur that do not at any other point of the scalar spectrum. Thus while the Eames Office flirted with the idea of shifting their scalar lens to the eyeball at 10^{-2}, they were ultimately drawn back to the skin.

This chapter asks if, why, and how certain scales are special. How, if you like, does cosmic-zoom media, concerned as it is with the arrangement of scales into a continuous medial assemblage, privilege certain scales above others in its arrangements? How does it reveal or obscure particular interconnections between scales and resulting scalar clusters? What, in short, are the politics of scalar constellation?

The Eames team created the "transparency clause" in their narration so they could keep the skin in the picture at 10^{-2}, but make it disappear at 10^{-3} and thereafter. Soon after, however, they brought physicist Philip Morrison and his wife Phylis on board and gave them control of the script. One of the first things the Morrisons did was to remove the transparency clause. The remainder of this chapter explores the clashing scalar politics of the

Morrisons and the Eameses, and the contradictory model of the universe that would emerge from this struggle as the most famous film ever made about scale: *Powers of Ten*.

UNIVERSAL MODULES

Philip Morrison was profoundly shaped by two trans-scalar events. The first occurred when he was three years old: the poliomyelitis virus penetrated his skin and left his body crippled for life. Thereafter he would require a cane to explore his environment, and later a wheelchair. According to Morrison, this gave him "a wonderfully enlarged view" by inclining him to "look past my confined self, first through books, then via radio, and among the wonders of the basement workshop bench."[10] The second was his involvement, as a young engineer turned physicist, in the Manhattan Project and its aftermath. Morrison worked on the detonation system for the plutonium bomb, helped assemble the Trinity test bomb as well as the bombs dropped on Hiroshima and Nagasaki,[11] and later directly assessed the damage at Hiroshima, an experience that would catalyze his transformation into a leading postwar critic of the nuclear arms race.[12] From atomic fission to the decimation of a city, Morrison's experiences already encompassed a wide range of scalar difference. But he would go further, studying cosmic ray propagation in the universe and the role of the neutrino, a massless particle. As we might expect, however, a mind tuned to the dynamics of both the smallest particles and largest galaxies found less pleasure in disciplined academic science than in popular science writing. There Morrison explored the multiscalar nature of the universe with increasing wonder. Everywhere he found the unfathomably tiny tied inextricably to the unfathomably large.

In one essay, "The Actuary of Our Species," Morrison assumes the long view of the insurance actuary who must assess the existential risk of the human race. What are our long-term chances of survival? Pretty good, it turns out, due to the spatial distribution of the species—the scale of its milieu—and the sufficiency of only a small community of survivors. At the same time, however, the potential scale of human exploration is relatively small vis-à-vis the universe. "I doubt much that we will turn the galaxy into a park, or visit the ends of the cosmos. It costs too much, in energy, time, and plan."[13] In the same way, the galactic politics of scale suggest that the universe is too large to meet other intelligences in the flesh, but the speed of minuscule particles means, according to Morrison, that interstellar communications are both likely to occur and likely to be effectively unidirectional: traversing the vast distances of space entails the unavoidable cost of a vast timescale.[14] This realization led him to establish the organization and set of strategies that would eventually become the Search for Extra Terrestrial Intelligence (SETI), which continues to systematically search for radio patterns originating in other star systems.[15]

At the atomic level, Morrison wondered at a further bifurcation of scale: atoms are decomposable but absolutely uniform (contain no individual variation from their type), while larger-scale objects are never uniform.[16] As we will see below, Morrison finds these intertwined scalar asymmetries to be key to the functioning of the universe. Scale, for Morrison, is not merely a set of size domains or a representational tool, but a fundamental characteristic of the universe, "down to the atom and out to the stars, and really prior to both."[17] What does it mean for scale to be prior to atoms and stars, or the objects that we measure using this conceptual schema? Morrison is interested in dynamic relationships and continually emphasizes the simultaneously structured and chaotic nature of motion in the universe, "the fusion of cause and chance, seen over the whole of physical phenomena."[18] Crucially, it is scale that provides the differential dynamic that ties the two together, that articulates the serial encounters of causation with the radical deviations of chance that generate new configurations. As both are necessary to produce structure at all, the differential of scale is a prerequisite for form, and in this sense is prior to any objects that emerge from material interactions. As we shall see, scalar differentials are the key to this articulation: atoms behave differently from stars, even though the latter are composed of the former.

By 1968 Morrison's professional interests had stabilized around explorations of scalar dynamics to such an extent that when he was invited to London to give a lecture series at the Royal Institution, he chose "Gulliver's Laws" as his subject.[19] That same year, he provided some consultation to the Eameses for *Rough Sketch*. When the Eameses approached him again in the mid-1970s to help with the new version, Morrison jumped at the opportunity, quickly taking a lead role as primary consultant, screenwriter (for the second draft), and narrator.

Morrison altered the *Powers of Ten* script in small but significant ways. At 10^{18}, as multiple stars are becoming visible, he added to the narration, "Who knows? These may be home skies to other planetary beings."[20] At the point where the virtual camera penetrates the skin of the picnicker's hand, he deletes the transparency clause, replacing it with "We will enter there . . . crossing the outermost layer of dead cells into a tiny blood vessel within."[21] The ellipsis in the script corresponds with the transition between

10^{-2} and 10^{-3}, and thus opens up a linguistic lacuna that contains the problematic shift of surfaces within its zone of indistinction. On one side of this shift we "enter" the body, on the other we are "crossing" the "dead" cells that mask the living interior. Morrison's solution to the problem of surface discontinuity, then, is to perform a discursive shift in perspective, from that of an observer "outside," in an environment with infinite potential vectors of movement, to one "inside," where a single vector directs us through successive layers, each closer to the vital center of the body, radial source of life itself.

Working only with a script indicating discrete scales, not yet with a film (he is not a filmmaker), Morrison has no conceptual difficulty introducing such a radical shift. For him, the world of the small is both continuous (in the sense of being always already "underneath") and discontinuous (in the sense of possessing radically different properties and capacities for interaction) in relation to the world of the large. Morrison's insistence on the radial structure of the atomic scale, his emphasis upon the perfection and centrality of atoms in contradistinction to the "traces of interaction" and broken symmetry of structures at all other scales, is buttressed by his final script changes. At 10^{-15}, the final scale represented in the script, the Eameses' original text declared that "we are surrounded by quark activity. Quarks have no mass (only charge and perhaps spin?) yet they seem to be, so far, the ultimate components of all the varieties of matter—and of energy."[22] Morrison replaces this description of a dynamic environment of energetic entities with a hesitant (if not downright skeptical) question: "Is this a bag of quarks in intense interaction?"[23] Morrison changes the subject of his sentence from "quarks" to "bag," focusing the viewer's attention on the container rather than the contents, resolving quarks as a singular set rather than a potentially boundless multiplicity.

Concomitantly, Morrison shifts focus from the ultimate to the penultimate scale: 10^{-14}, the width of the nucleus of the carbon atom. "We are in the domain of universal modules," writes Morrison exuberantly. "There are protons and neutrons in every nucleus, electrons in every atom, atoms bonded into every molecule, out to the farthest galaxy."[24] Morrison privileges this scale for the same reason that Democritus, writing twenty-four centuries earlier, privileges *his* scale of the atom: it is the scale that makes all of the others possible. For Democritus, the atomic scale is virtual, as he could not measure or observe the size of the *átomos* that he posited. But these uncuttable forms, whose shapes determine the potential structures of all larger assemblages, presented for Democritus a scalar structure of homology. "Democritus in assigning a shape to each quality made sweet to consist of fairly large, spherical atoms. To the quality sour he assigned very large, rough shapes with many angles and no curves. . . . Salt is angular, fairly large, twisted, although symmetrical."[25]

Morrison's atoms are different in several respects, including their de-composability (into protons, neutrons, electrons, and eventually maybe quarks) and their structural *discontinuity* with the macro forms into which they assemble. Thus Democritus's scale-homologous, uncuttable, imperfect atoms are replaced with Morrison's nonhomologous, cuttable, perfect atoms. Where structural homology stabilized the scalar continuity in Democritus's discursive system, Morrison achieves this effect through an ingenious coupling of atomic-scale perfection with acknowledged discontinuity. "Two worlds remain," Morrison writes elsewhere, "the world of the shimmering atoms, and the world of human senses, telling us something about averages. But the finest instruments cannot do better; there, too, immediacy and directness are but illusions of familiarity." The radical, experientially unbridgeable discontinuity between these two worlds secures the unfathomable, unbreachable perfection of the atom, cuttable but not corruptible, an entity upon which collisions and other interactions leave no trace, in contradistinction to larger entities like solar systems, which "bear the marks of every previous collision."[26] For Morrison, then, atoms and their primary components can have no history, no record of change. Time encoded in space as change is a phenomenon that emerges only at larger scales. Yet it is the atom's purity that undergirds it all. This is what authorizes Morrison to speak of these entities as "universal modules." Discursively, his universality signifies both a scalar common denominator and a stabilization of causality itself, a double buttress against chaos. Medially, his universal modules become a collective scalar node within *Powers of Ten*, with sinuous links to every other scale depicted in its frames. 0.0001 angstroms: Morrison's scalar center of the universe.

Morrison's scalar politics are essentially binary: the scalar spectrum is divided into two halves, the micro and the macro. The micro is without history, without change, while the macro is always in flux, suffused with events. For Morrison, then, the atomic scales are privileged. It is clear that he prefers the beauty of timelessness to the vibrancy of macro chaos. The scale of skin (10^{-2}) isn't particularly significant to him, and the minor barrier it presents to an optics of the zoom is entirely inconsequential. In *his* scalar spectrum, things get more and more exciting as they approach the atomic scale. Another consequence of this particular scalar privileging, however, is that scales smaller than the atomic make him uneasy. Particles that contain only one-third of a charge? That are massless, and yet are the constituent parts of the atomic nucleus? That are decomposable entities, yet cannot exist independently? That are completely indeterminate with respect to position, velocity, and trajectory? That may wink in and out of existence? None of this sounds like the serene perfection of timeless beauty that Morrison sees in the atom, and accordingly, he deemphasizes the scales of the quarks. He discursively figures quarks as something inside a bag, a bag intended to contain their indeterminacy, their shifting qualities, and potentially their *history*. Quarks are unruly, and the figure of the bag suggests a strong scalar schism below the atom, just like the strong scalar schism above that marks the boundary between Morrison's "two worlds." Here is a third world, an abyss of unruly, partial beings. The lower scalar bounds of the atomic hold this chaos at bay by acting as a container, a scalar membrane that converts a wriggling mass of imperfect partial particles into a stable, unchanging entity that makes use of these smaller parts, but only by regularizing them, rationalizing them, and permanently fixing their characteristics.

We may laugh at the fact that Morrison feels so strongly about atoms and the impingement of quarks on their pristine beauty. But scalar politics reign at all scales, and I am quite certain that you, the reader of this book, entertain your own scalar biases. The Eameses, as we have seen, were deeply invested humanists. While their scalar spectrum may look stripped of distinction, rolled into one endless surface, their philosophy of design and their use of the medial apparatus as a surrogate for the human subject ensure that one scale is held steady even as all others zoom past. While *Rough Sketch* and *Powers of Ten* are both figured as zooms, I have argued that medially they are more like loops, modeling the universe as a scalar continuum that is always equidistant from the viewing subject. This is what pan-scalar humanism looks like as a scalar ideology.

What happened, then, when these two radically different scalar ideologies clashed during the making of *Powers of Ten*? In the final two drafts of the script, the Eames team eliminated Morrison's reference to alien beings inhabiting the stars, and let the quarks out of the bag, albeit still without the benefit of existential assertion ("Are these some quarks in intense interaction?"), but kept Morrison's unapologetic penetration of the skin as well as his universal modules. The Eameses, as we have seen, were primarily concerned with temporal and spatial continuity and a universalized, multiscalar vision stabilized by iterative computation. Morrison's universal modules, by connecting all scales in a virtual network, seem to have gelled with this project of continuity, despite his more nuanced view that this universality is purchased at the expense of introducing a radical

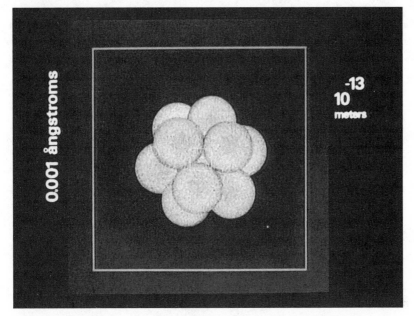

Figure 15 ▸ "We are in the domain of universal modules." Frame from *Powers of Ten* by Charles and Ray Eames (1977).

scalar break between the atomic scale and all others. The Eameses transposed Morrison's networked material universalism into the form their 1968 film had assumed: a hierarchical ladder, looped around the human observer. In order to do this, their team had only to eliminate Morrison's emphasis on the scalar discontinuity authorized by quantum mechanics, leaving half—and only half—of his story intact. This discursive transformation was achieved not through direct omission but through addition: primarily, the addition of the iterative, discretized, but inexorable temporality of the filmic medium itself, and of the continuous zoom, both of which they (along with Bronowski) had innovated in *Rough Sketch*. In the resulting visual sequence, Morrison's "universal modules" narration is edited so that it spans two scales (10^{-13} and 10^{-14}), during the approach of the carbon nucleus (figure 15).

While Morrison had striven to place the primary narrative emphasis on the universal modules at 10^{-14} meters, the Eameses were concerned to emphasize the human scale and the absolute distance (measured in powers of ten traversed) that they were able to visualize. Thus they zoomed right past Morrison's universal modules, into the realm of massless quarks, and insisted that the narration end with the impressive number "ten to the fortieth, or one and forty zeros."[27] Taken out of the context of Morrison's

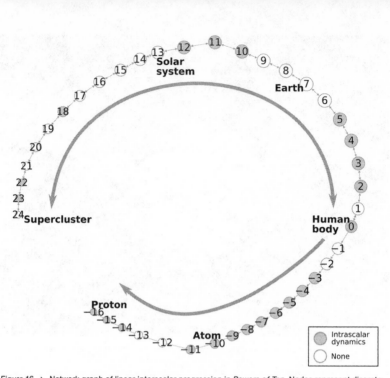

Figure 16 ▸ Network graph of linear interscalar progression in *Powers of Ten*. Nodes represent discrete scales, edges (directional arrows) indicate interscalar connections, and shaded nodes indicate scales that contain intrascalar dynamics, or events occurring within their domains.

other writings, his claim about universal modules could be interpreted to support rather than disrupt the hierarchical, linear arrangement of scales produced by the film. Universal modules can be arranged hierarchically as well as radially, and the discrepancy between the narration's implied scalar structure and the film's explicit one has, to the best of my knowledge, gone unnoticed—or at least unremarked—by scholars. These structural differences can be easily seen in the context of a network graph, however. If we chart the interscalar relationships in the film as a linear unfolding (figure 16), which is to say, if we only consider *Powers of Ten*'s continuous, *visual* scalar connections, it quickly becomes evident that within the film's linear temporality, some scales are emphasized by virtue of the fact that they appear twice (once in each direction of travel, indicated in the figure by bidirectional arrows), but no scale is fundamentally connected to any other beyond its spatially contiguous neighbors.

If, on the other hand, we produce a network graph that takes into account not only the *visual* relationships between scales but the discursive, *textual* relationships established by the film's narration, a multidirectional, highly interconnected network becomes visible. Visualizing *Powers of Ten*

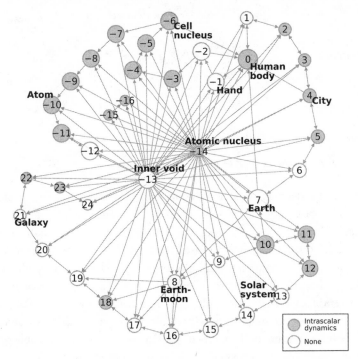

Figure 17 ▸ Network graph of interscalar references in *Powers of Ten*, arranged by centrality.

as a map of the totality of its interscalar references (figure 17) reveals the discursive centrality of the smallest scales in the film's speculative scalar ecology. Via Morrison's narration, these connections fold the film inward, producing a cohesive interscalar space that includes but extends (dimensionally) above the linear plane of its spatial continuum. In this way the film replicates some of the interscalar potentials of Boeke's *Cosmic View*, which contains a similar networked structure within its smaller scales.[28] While Boeke's book possesses only a linear set of spatial connections between its larger scales, *Powers of Ten*'s universal modules are discursive "super nodes" that possess outgoing links to all other scales, thus drawing them into a network with a shorter average path length (the maximum distance from one node to any other node) of 7.35 versus *Cosmic View*'s 10.75.[29]

My implicit argument in representing *Powers of Ten* as a network, which will be developed further as we encounter later works, is that scalar media possess interscalar "shapes" as well as sets of intrascalar dynamics. Together they define the possibility space of scalar interactions. These network graphs, then, capture an essential dimension of the mediation that these works perform. For instance, the atomic scales (10^{-13} meters, the space

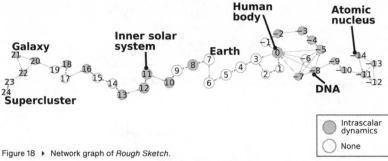

Figure 18 ▸ Network graph of *Rough Sketch*.

between the electron cloud and the nucleus of an atom, and 10^{-14} meters, the nucleus itself) occupy absolutely central positions within this network. These scales connect to all others, which means that in navigating the scalar spectrum (conceptually or otherwise), the most efficient path between most pairs of scales, no matter how radically disparate, will pass through them. At the same time, however, certain other scales are privileged as sites of intense intra-action, or activity within their discrete scalar milieus. The region of the scalar spectrum that is most vibrant, in this model, is that between the atomic scale and that of the multicity region (10^5), with a few notable gaps, as between the outer electron shell and the nucleus of the atom.

Compare this with a network graph of *Rough Sketch*, rendered using the same algorithm (figure 18). The linear structure of its interscalar connections is immediately visible, but so are a few clusters of local interconnectivity between scales: the galactic (10^{20} to 10^{22}), the cellular (10^0 to 10^{-8}), and the interatomic (10^{-11} to 10^{-14}). The human body, as in the later *Powers of Ten*, remains relatively central and interconnected—however, it must be noted that the human body has more incoming than outgoing connections. This suggests that the body is affected by more scales than it affects in turn, through the expression of trans-scalar processes such as DNA transcription, the collective functioning of cells, and so on.

This graph also indicates that although the work of the Morrisons resulted in a far more elaborate narration, and therefore many more articulated interscalar connections, the later film did not break with the basic interscalar structure of *Rough Sketch*. *Powers of Ten* grows many edges in comparison to its earlier version, but most of those edges spring from the same privileged regions of the scalar spectrum. If we consider this clustering alongside intrascalar dynamics (the shaded nodes on the graph), we notice something strange. While only twenty-two out of thirty-nine discrete scales in *Rough Sketch* are portrayed as containing internal interactions (i.e., dynamic processes of any kind), those active scales tend to be located in exactly these clustered regions. This isn't the case in *Powers of*

Ten, where dynamic activity gravitates toward the smaller end of the scalar spectrum. Ecologically, as a model of the cosmos, *Rough Sketch* produces a set of doubly privileged scales: those with internal dynamics are also interconnected, while "lifeless" or inert scales are also relatively isolated. The interior of the human body, the inner solar system, and the Milky Way galaxy are the privileged scales in this ecology, the vibrant regions of the scalar spectrum.

I return to comparative readings of cosmic zooms rendered as interscalar networks toward the end of this chapter. It is important to remember, however, that scalar relationships can be embodied in a medial work in very different ways: visually contiguous space and textual narration produce differently legible links for different viewers. Figure 17 represents both sorts of connections, figure 16 only those that are visual. As we can see, the difference is dramatic. This, I believe, at least partially explains the seemingly harmonious coexistence of two radically opposed scalar ideologies: Morrison's conception of the film as an expression of universal modules stabilizing all scales simultaneously and the Eameses' conception of it as an expression of frictionless human access, a model of a linear but continuously scalable horizon. The complex interplay of text and image, language and motion, creates a multilayered medial system "thick" enough for both conceptual structures to find discursive support, but it should be clear that the linear, visually reinforced structure represented by figure 17 obscures and attenuates Morrison's to a significant degree. Nonetheless, both potentials exist to be extracted by the differential attentions and efforts of different viewers.[30] This is one reason to perform critical readings of culturally influential cosmic zooms. Any cosmic zoom is a means of modeling the universe and its potentials: what encounters it kindles, and what avenues it forecloses.

The scalar politics of today are waged in media that construct subtle cosmic zooms, competing ecologies that privilege particular scales as vibrant regions of the scalar spectrum. As certain cosmic zooms achieve relative dominance within a culture, their scalar biases begin to structure the very ecologies they purport to describe. As Ursula Heise argues in *Sense of Place, Sense of Planet*, it is precisely these biases (particularly in favor of "local" scales, or places) that have stood in the way of an environmental politics adequate to the risks we face.[31] Which scales should we privilege? And how, as a point of rhetorical strategy, can we shift or distribute care from one scale to others by constructing implicit scalar networks? The answers to these questions will largely determine the course of the Anthropocene. The battle to resolve planetary and geological scales is not a battle to shift scalar focus (to the earth, for instance), but rather to construct a cosmic zoom that articulates the right scalar network, tells an effective story of interscalar (and intrascalar) dynamics, and enables the imagining of new processes in previously devalued regions of the scalar spectrum. Scalar politics is a politics of the network.

TECHNOSCIENTIFIC OPTICS: FROM CONTINUITY TO CONTIGUITY

Perhaps the most significant aesthetic choice during *Powers of Ten*'s production, the drastic alteration of the "instrument panel," tipped the scales decidedly toward the smooth model of scalar contiguity. One contributor to the internal debate (presumably Alex Funke) proposed that the elaborate instrument panel from *Rough Sketch*, predicated upon the motif of travel, be replaced with a simpler gauge, resembling a "thermometer," that would depict field of view (given, as before, in powers of ten meters) as a scalar hierarchy. An "arrowhead" would move up and down its axis as the viewer navigates the film's scalar hierarchy. The author of the proposal notes that this format "allows one to circumvent the time dilation problems entirely" by suggesting that the changing field of view relates not to a virtual traveler but to a virtual observer "with a zoom camera situated a great distance from the earth . . . the distant observer simply turns the zoom lever on his camera by repeated factors of ten at whatever pace is comfortable to the viewer: no clocks are displayed at all."[32] This scheme, expressing the fantasy of a cosmic-zoom lens, would have better correlated with the actual drama of resolution enacted by both Eames films, as discussed above. Nonetheless, it was not implemented. Philip Morrison, joined by his wife Phylis as consultant for the film, ending up making a similar suggestion: keep the powers of ten indicator (still motivated by a motif of travel) but "drop" the relativity clocks.[33]

The Eameses agreed with the Morrisons, and the clocks were dropped (see figures 14 and 15). This allowed the filmmakers to dismantle the "dashboard" altogether. To the left of the square image, the length of a side, and the viewer's supposed distance from the scene, is spelled out (from "1 meter" to "100 million light years," then down to "0.000001 ångstroms"); in the upper right corner, the measure is repeated, in meters, as a power of ten (from 10^0 to 10^{24} to 10^{-16}). The final film, then, represents a hybrid of *Rough Sketch*, the understanding of scale brought to the project by the Morrisons, and the production team's suggestion to embrace a conceptual zoom rather than a travel motif. By reformulating the film's interface with the viewer to resemble a shifting viewpoint with superimposed, supplemental information rather than the output of a spaceship's instrument panel, one level of conceptual and aesthetic mediation is rendered opaque. No longer sutured to the subject position of a virtual astronaut, the viewer no longer relies upon a spacecraft to speculatively travel to other scales; rather, she gains the capacity of radically extended vision, a cybernetic coupling with a zoom lens. This expanded vision is relegated to a linear axis, the hierarchical axis of the film's drama of resolution, but the ability to move up and down this hierarchized set of resolutions now comes without any relative temporal cost. That is, by removing the relativity clocks, along with any mention of relativity in the narration, the filmmakers have rendered the temporality of their simulation absolute. Because the hypothetical observer's time is now identical to the medial temporality of the film itself (24 frames per second), and thus to the viewer's experience of the film's unfolding, the effect of removing both clocks is to eliminate the reference clock on earth and stabilize the remaining interlocked temporalities as an absolute frame of reference.

The removal of any temporal reference to the opening scales of the film, and thus any differential between the perspective of the viewer and the perspective from other scales, constitutes a shift toward occluding the material grounds of the film's continual scalar negotiation. That is, the reference (virtual) surface that this medial system resolves onto a medial surface (screen) is no longer framed as constrained by the same physics that govern the existence and continuance of the latter surface. Thus the medial act in which the viewer participates is relocated to a virtual space outside of the environment that it references. For Morrison, this may have seemed like an appropriate analogy to the freedom of becoming, the "chance" that suffuses the universe. For the Eameses, it shifted the focus of the film from simulating a trans-scalar spaceflight to simulating a timeless (optical) resolving of the universe at every scale. The paradigm shifted, between *Rough Sketch* and *Powers of Ten*, from the kinetic mediation of the Apollo program—technology as vessel for human exploration—to the visual mediation of technoscientific optics: knowledge as disembodied. This conceptual shift is symbolically represented in the change of location from Florida (site of the Apollo launches) to Illinois, in the central region of the North American continent. Members of the Eames Office later noted that the geographical shift was inaugurated "to allow the journey to approach the disk of the galaxy approximately at right angles."[34] What mattered to the filmmakers was not the perspectival divergence of *Rough Sketch*, but rather a convergence of disciplinary scientific modalities of knowing, from physics to biology to astronomy.

With the removal of relative temporality from the film's semiotic register, the nature of the cosmic zoom's continuity changes. In *Rough Sketch*, as we have seen, the smooth zoom produced at the viewer's timescale is purchased by continually varying the temporal scale of the virtual environment being imaged. In *Powers of Ten*, this continuity of resolution and field-of-view change is disarticulated from the environment itself. Rather than negotiating a relationship between two diverging positions within the same virtual environment, the medial assemblage articulates the reference surface to the position of the observer (and thus viewer). The resolution of successive scales becomes an extra-environmental event, unfolding at the viewer's timescale, but not in accordance with the dynamics that pertain between two surfaces within the viewer's universe. While the first film establishes a scalar *continuity* between an environment and a spacefaring traveler that diverge in time, the second establishes instead a scalar *contiguity* in the form of a continuously unfolding surface that parallels the viewer's own.

In an undated, handwritten document that appears to have been produced early in the preparation of *Powers of Ten*, one of the Eameses' researchers compiles a list of "new material, not available last time round," thus articulating one of the primary rationales for producing a new version of the film: more artwork had become available to produce a seamless composite. These new materials included astronomical output from the Palomar Observatory, scanning electron micrographs from the 10^{-3} to 10^{-5} meter scales, a macro-scale model of DNA, and "satellite photography in God's plenty."[35]

Most scholars read *Powers of Ten* as merely an updated version of *Rough Sketch*, when they mention the previous film at all. Eric Schuldenfrei, in *The Films of Charles and Ray Eames*, for instance, glosses over the differences between the two films with this succinct argument: "The final 1977 version eliminated the superfluous elements and centered the main frame. Through progressive versions of *Powers of Ten* the Eameses improved the film by reducing the information shown."[36] A 1989 book coauthored by Ray Eames simply notes that "in the ten years between the two versions, many advances in theory and research had occurred."[37] Sylvie Bissonnette, in her survey of "scalar travel documentaries," with *Powers of Ten* as its chief object, mentions *Rough Sketch* only obliquely, noting that the later film "is an expanded version of an earlier sketch made in 1968."[38] Conversely, I have been arguing that *Rough Sketch* is profoundly different from *Powers of Ten*. This divergence extends to the latter's approach to image plates as well as its approach to the "dashboard," narration, and characterization of the scalar spectrum. *Rough Sketch* made use of photographs of Florida, skin, and the whole earth, smoothly blended with the drawings that make up the majority of the film's simulated milieus. The sources of the images were unimportant: the film's principle innovations lay in its ingenious production of a relativity simulation based upon iterative powers calculations that stabilized a continuous zoom through all known scales of the universe. *Powers of Ten* abandons this approach in favor of a simplified simulation of a single contiguous surface. Photorealism thus becomes the dominant aesthetic, in support of which the Eames Office searched long and hard for actual photographs and micrographs for as many scales as possible. These didn't function as mere upgrades to the previous artwork; they were deliberately placed in a central aesthetic position by the decisions to emphasize universality rather than difference, and space rather than time. The satellite imagery enthusiastically noted to be available "in God's plenty" stands in metonymically for all scale-mediating visual technologies that fix a relationship between an observer and an observed surface. Like manna, these are the technologies *of* the gods. A lavishly produced sales brochure for the film, meticulously designed by Funke and the Eameses, makes it clear that the film performs a synthesis of visualization techniques used in the sciences:

> Along one unbroken exponential path, *Powers of Ten* draws upon the latest available images from many technical and scientific disciplines: classical and radio astronomy, new maps of the galactic clusters and the Milky Way, large-format aerial, mapping, and satellite photography; optical, electron scanning and transmission microscopy; x-ray diffraction analysis; and new models for those subjects that still lie beyond our vision.[39]

Each of these visualization techniques comes with a set of disciplinary protocols, including the particular scales that they stabilize through the twin acts of capture and spectatorship they enable. The film's contribution is imagined to be connective—suturing these disparate visualization techniques and their corresponding scales into "one unbroken exponential path." The images, then, serve as more than mere plates to be utilized on an animation stand to produce a certain effect; they are the subject itself: man's capacity for mediated visualization.[40] Here the "vision" evoked in the brochure's final sentence does not refer to the capabilities of the human visual cortex, for nearly all of these scales lie beyond them. It can only mean the technoscientific enterprise of *visualization as a whole*, a type of seeing that is both mediated and mediating.

The technoscientific optics alluded to here conjoin two organizational principles or vectors of knowledge production: intensity of the frame and extensity of the zoom. As I've argued throughout this book, the zoom does not increase detail. Rather, it produces a continuous surface equidistant from the constructed (viewing) subject; it is thus better thought of as a loop. Another way to put this: the type of zoom enacted in *Powers of Ten* is not intensive—that is, it does not measure or reveal difference contained within a single surface. Instead, it blends different surfaces (scales, rendered as image plates) and organizes them around an implied subject. This is an extensive medial process because it continually unfolds a linear chain, theoretically without end. How large can we go? How small? Derek Woods even suggests that this linear unfolding through scale domains has the capacity to give the viewer access to "epistemic things" that are real but not yet entirely assimilable to thought. In these cases, "the trope does not so much domesticate nonhuman scale domains anthropocentrically as move an epistemic thing from a domain of prosthetic visibility to one accessible only to complex stabilizing apparatuses."[41] In these cases, adjacency on the scalar spectrum, when performed by cosmic-zoom media, can be mobilized transitively to nudge established knowledge and newly discovered forms of difference closer together. We can also add, as one of the complex stabilizing apparatuses that solidify an epistemic thing from a radically different scale into that of the human: the blue, superimposed bounding boxes contained in both *Rough Sketch* and *Powers of Ten*.

Like matryoshka dolls, the pale blue boxes of the Eameses' films contain scales within scales. These boxes are a mutation of those in *Cosmic View*, but their animation here changes their meaning. They are no longer merely traces of previous acts of resolution, of the scalar negotiation between two surfaces set up by Boeke's medial system. In the Eameses' films, these boxes become a continual nesting, in principle without end and without necessary reference to any previous act of resolution.

Each of these boxes defines a particular scale, an aesthetic effect that interrupts the smooth zoom, reminding us that the zoom is not a set of scales, but the ligaments *between* scales (arcs of resolution). Individual boxes, superimposed over their realistically rendered environments, remind me of quadrats. A quadrat is a device—conceptual, but also often physical—used by ecologists to delineate an area to be empirically studied. The quadrat superimposes a bounding box over a slice of an environment, providing boundaries for specimen counting. A field ecologist seeks to exhaustively count all specimens contained within her quadrat, and none beyond its borders. As a standardized frame, the quadrat aids quantification by providing a reference area, required to calculate density. One of the challenges of quadrat use, however, is that they can be constructed at any scale, and there is no standard by which to determine which quadrat to apply to any particular study. This has led to periodic critiques from within the field of the often arbitrary scale decisions made by field researchers, whose sample scales "tend to reflect hierarchies of spatial scales that are based on our own perceptions of nature. Just because these particular scales seem 'right' to us is no assurance that they are appropriate to reef fish, barnacles, anoles, cattle, or birds."[42] I have argued elsewhere that the scale-relative nature of ecology is not a limitation but a call to trans-scalar discovery.[43] As J. A. Wiens implies, the question of proper scale is one that can only be answered by disrupting our anthropocentric scalar models and learning what scales are relevant to other beings and processes. I further expand on the connection between ecology and scale in the next chapter.

In *Powers of Ten*, the quadratlike boxes both stabilize particular scales as potential knowledge-producing milieus and mark the inexorable disappearance and replacement of any stable milieu. In Boeke's *Cosmic View*, the equivalent boxes indicate a determinate material relationship to previous pages in the book. In the Eameses' films, the powers of ten become transcendent, standing in for the names of scales, bounded by their respective boxes. These blue boxes indicate, perhaps unwittingly, the predefined grid of knowledge-making that undergirds this project of pan-scalar humanism. The boxes, instrument panel, and narration all reinforce the external position of the human observer, whose mastery is almost mystically derived from an original one-meter cube.

As we have seen, Denis Cosgrove refers to this form of seeing as Apollonian, a simultaneous quest for transcendence and mastery. "The Apollonian gaze, which pulls diverse life on earth into a vision of unity, is individualized, a divine and mastering view from a single perspective. That view is at once empowering and visionary, implying ascent from the terrestrial sphere into the zones of planets and stars."[44] Indeed, the earth makes a curious appearance in *Powers of Ten*. According to the rules set up at the opening of the film, a blue box will appear at each successive tenfold increase or decrease in field of view (FOV). Given the one-meter-square starting position, this yields an image of the whole earth a bit after 10^7. That is to say that 10^7, or ten million meters, is somewhat less than the diameter of the earth, and the result, as seen in *Rough Sketch*, is that the sphere of the earth exceeds the blue scalar box that marks the boundaries of that particular FOV. In *Powers of Ten*, which duplicates both *Rough Sketch*'s FOV assumptions and reference scale, this blue box is miraculously shifted so that it perfectly encompasses the earth's diameter, neatly pack-

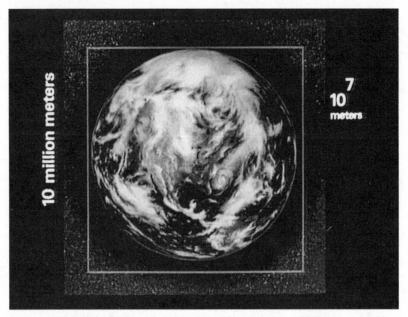

Figure 19 ▸ Packaging the planet as object of technoscientific control. Frame from *Powers of Ten* by Charles and Ray Eames (1977).

aging the planet, as it were (figure 19). This colossal cheat would serve no purpose in *Rough Sketch*, concerned as it is with the powers of ten as enabling iterative technique. Ironically, however, the IBM-funded version of the film demotes the precision of power thinking in favor of literally and figuratively framing the photographic image as aesthetic object and expression of scopic prowess.

At the same time that *Powers of Ten* marks its constituent imagery as something made, as an expression of technologically enabled scalar mediation (each artwork plate stabilizes a scalar relationship to the viewer), it erases its own self-reflexivity. This, in my view, is one of the central paradoxes of the film: its own technique is rendered opaque even as it frames and exalts the production of technoscientific imagery of radically disparate scalar environments. *Powers of Ten* is the least self-reflexive of all the instantiations of the cosmic zoom we've discussed thus far. *Cosmic View* reflected upon its own techniques and constraints of production at no fewer than thirty-nine scales. *Rough Sketch* did so at eleven scales. *Powers of Ten* reduces this to a mere four: between the scale of the earth and of the solar system, at the galactic scale, and at its largest scale, near the outer edges of the universe.

Kees Boeke, when faced with resolving the universe as a whole, found the task impossible. In place of an image at his scale 27, he provides only a textual explanation of the limits of representation:

> On this page there might have been put a drawing 27 to continue the series, representing what would be seen if we could take up a point of observation ten times farther away . . . [but] it would be quite impossible to draw the galaxies and clusters of galaxies small enough and near enough to each other. But also, the limits of what is supposed to be the curved space of our universe would be within that 27th square, and there would be no possibility at all of portraying or even visualizing the "curvature of space," which would be the determining factor there.[45]

Boeke acknowledges both the material limitations of resolution and the representability of space itself. This scale, the scale of the universe as a whole, cannot be stabilized within his medial system; any such negotiation is impossible. A system with a vastly enlarged resolving power might be able to solve the first problem (though it would likely run into the resolution limits of the human eye), but no medial system that relies on a visual modality of scalar negotiation will be able to bridge the radical alterity figured in Boeke's text. Thus Boeke *does* partially stabilize this scale (one billion light years), but not within a visual medium.[46] By contrast, *Rough Sketch* ends one power earlier, at Boeke's final image (one hundred million light years), but suggests that this is an arbitrary stopping point, that we could in principle continue to zoom out: "Were we to continue our journey, the scene would probably now remain much the same. The spots of light might just get smaller."[47] Thus the Eameses reduce Boeke's representability limitation to his resolution limitation, and even then suggest that it isn't a real problem: only the danger of boredom keeps us from going farther. In *Powers of Ten*, the film actually passes beyond this scale, partway to a billion light years, but stops well before reaching it. "As we approach the limit of our vision, we pause to start back home," narrates Morrison laconically. Thus, while *Powers of Ten* acknowledges that a limit has been reached, it remains, like *Cosmic View*, cryptic about the nature of this limit: is it one of resolution, of fundamental representability, or of contingent visualization technique? By holding open the latter possibility—that is, that "vision" is being used in the sense of "the technoscientific enterprise of visualization," as in their sales brochure—the film produces a continuing virtual path like that of *Rough Sketch*, stretching in principle farther and further beyond the universe itself. We, as spectators, only "pause" at this contingent limitation; at the other side of the pause lies not only the imminent inward journey, but also the inevitable march of technoscientific knowledge production within a visual modality, ever gnawing at the limits of our transcendent view of matter.

In "The Age of the World Picture," Martin Heidegger suggests that the metaphysics of the modern age is characterized by an ordering of the world into a coherent totality that drives scientific research as a never-ending, ongoing activity. "Wherever we have the world picture, an essential decision takes place regarding what is, in its entirety. The Being of whatever is, is sought and found in the representedness of the latter." Heidegger's key point is that the essential representability of the world precedes any research into it; science thus discovers that which it has been primed to see, the conditions of which are the disarticulation of the human from the vicissitudes and discontinuities of the world in order to stand outside of it, as its condition of representability. "Man becomes the relational center of that which is as such."[48] Knowledge, in this mode, becomes a projection of a world picture onto the world, which takes the form of scientific research. Thus, the picture of the world isn't something that is intuited, that comes together over time as a result of experience. Rather, the world as picture is a starting point that subsequently directs all activity of knowledge production. Certainly with regard to scale, *Powers of Ten* exemplifies such a world picture. Though it explores only a thin slice of the world, it produces a scalar continuum that enacts at the same time a total representational system (anything found within the totality of the world exists somewhere along its scalar axis) and a contiguous field that eliminates any discontinuities in the ordering of the universe. *Powers of Ten* is thus an expression of a world picture, an illustration of the diagram of representational forces that bring it to fruition out of the age of technoscience. At the same time, however, it is a singular, concrete representation that characterizes and privileges particular scales, with a cultural provenance (traced in these pages) and future role, binding new thinkers—professional or amateur—to the world picture underlying modern technoscience.[49] Where Boeke had hesitated to declare the world's scales fully explicable as a continuous, totalizing picture, the Eameses and Morrisons aestheticize exactly this picture, enrolling new knowledge producers into its particular mediations of space and scale.

In its shift from an experiential temporal continuity (experiencing space as unfolding time) to a photorealistic articulation of a technoscientific contiguity of vision, *Powers of Ten* hit its stride, becoming likely the most influential science film of all time, and certainly the most celebrated of the Eameses' films. As we have seen, this was not the inevitable outcome of an intuitive representational process, but rather the result of decades of work by Boeke and two Eames Office teams, Funke's animation techniques and vision, and the narrative abilities and commitments of Philip Morrison, who spent his entire life exploring and reflecting on scalar dynamics. Perhaps the film's greatest irony, then, is that by incorporating Morrison's vision of matter as bifurcated between atomic incorruptibility and macro irregularity, between a realm where history is impossible and one where it is unavoidable, between causation and chance, the film mediates the viewer's relationship to a set of alterior scales that lack Morrison's fundamental division.[50]

The Eameses' use of an aesthetic of contiguity to subtend and unify discontinuity in the medium of film was not new to *Powers of Ten* or even *Rough Sketch*. They experimented with this basic strategy as early as *Glimpses of the U.S.A.*, created for the 1959 American National Exhibition in Moscow. Eric Schuldenfrei argues that in this work of propaganda aimed at a Russian audience they used Soviet montage theory in ways that produced continuities rather than disjunctures and dialectic juxtapositions, weaving together images disparate in time and place to construct one unifying vision of work and material abundance in the United States. "Between the cuts, performers dance, athletes run, and seasons merge into one another, from summer to winter and back to summer again, giving the still images a sense of continuous motion."[51] The film, composed mostly of twenty-two hundred still images, utilized seven twenty-by-thirty-foot screens.[52] It thus delivered both an overwhelming barrage of fragmented "glimpses" and a spatially contiguous overall vision of idealized capitalism. As I argue in the preceding section, *Powers of Ten*'s dual registers, once they are teased apart through scalar analysis, reveal a confused set of politics rather than a single, unified vision, even if one of its visions is that of homogeneous scalar contiguity.

As Val Plumwood has emphasized, the dominant mode of rationalism in Western culture has been a dualistic form of reason that "hyper-separates" the rational subject from the animal, the female, the passive, and the material—that is, "nature." At the same time, this rationalism denies its own dependencies (of the human on the animal, the mind on the body, the male on the female, etc.). It is thus a profoundly anti-ecological form of thinking: "Nature includes everything reason excludes."[53] The overview presented by scalar contiguity draws upon and continues this work of hyper-separation or radical exclusion by stabilizing a scale-free subject at the same time that it orders the world as a contiguous space rather than as a series of scales that cannot be resolved simultaneously. This contiguity is thus passive, inert, and non-agential: each scale is limited to interacting with its contiguous neighbors as if they were arranged on a single linear axis. This authorizes access for the rational mind, which divides this axis into discrete but arbitrary scales in order to study and produce knowledge from them. Rationalist, technoscientific discourse has made the resolving cuts, and highlights that fact. Its ideological tent pole is the claim that underneath these scalar boundaries, all is continuous and linear: matter behaves the same at all scales.

This is one of the central paradoxes of *Powers of Ten*: it produces a nested set of matryoshka-like scales, imposing blue boxes to indicate hard scalar boundaries, yet explicitly figures these resolving cuts as the products of disciplined knowledge production—the sciences. Underneath these disciplinary cuts, the film suggests, lies a unified scalar continuum. This, then, is the film's greatest ideological victory: it purports to explore the many scales of the cosmos while undermining the very notion of scalar difference. Scalar difference is the product of technoscientific technique, an arbitrarily articulated if consequential research program. In other words, scientific discipline produces scale in order to understand the underlying contiguity of the cosmos.

The central argument of this book is the opposite: scientific discipline is largely homogeneous, while the universe is fundamentally scale-articulated. This is not, incidentally, to denigrate the epistemic or pragmatic importance of science. Far from it, as I indicated in chapter 1 and will explore in the next chapter in particular, scientific discipline plays a vital role in the particularly human mode of trans-scalar encounter by making resolving cuts that open receptive pathways to the experience and knowledge of scalar difference. We might oversimplify by noting that culture posits contiguity and linearity, while matter produces its own differential cuts. While disciplinary resolving cuts are arbitrary, actual scalar difference is detectable and irreducible when we foreground our medial processes of resolution. In his own idiosyncratic way, Philip Morrison tried to indicate as much in his early drafts of *Powers of Ten*'s narration, only to be overruled by the Eameses.

Whether we consider *Powers of Ten* to have dramatically influenced or only to have reflected this particular ideology of scalar contiguity and frictionless epistemic access, its success is undeniable. Here the rational human subject, disarticulated from the processes that conjoin different scales and constitute the boundaries between them, reorders them as a contiguity and thereby positions himself as outside of scale itself. The subject, in this dominant narrative of Western culture, denies its own ecological embeddedness at one scale in order to gain epistemic access to all scales as contiguous knowledge domains. *Powers of Ten* draws upon and uniquely contributes to this particular narrative of mastery.

Scale, suggests the film, is smooth: we can slice it any way we like, but ultimately our Apollonian vision allows us to access its underlying continuity and articulate all scales together seamlessly. Bruno Latour has argued that this tendency, which runs counter to what we know about reality—"neither the schema of space, nor that of time, appear continuous: levels of reality do not nestle one within the other like Russian dolls"—arises when we confuse optical mediations with cartographic scale: "In spite of appearances, the optical and cartographic metaphors do not overlap. It might even be said that the former has become so parasitical on the latter that it has rendered the very concept of cartography almost incomprehensible."[54] As I have argued in these past three chapters, *Powers of Ten*, despite its provenance as an adaptation of Boeke's *Cosmic View*, marks the endpoint of a series of subtle transitions away from the scalar mechanisms of cartography in favor of a new aesthetics of the zoom.

As we shift away from the discrete negotiations of scalar difference staged by *Cosmic View* to the vision of scalar contiguity modeled through *Powers of Ten*, the tradeoff between field of view and resolution becomes not a negotiation of a single scale but a universal dynamic, a negotiation that need only be implemented *once* in order to deliver before us the entire menagerie of a contiguous world. Any conceptual barriers to establishing arbitrary relationships among scales are removed. In principle, this world can be navigated at will by the human observer. This, in my view, is the film's primary legacy. Millions of viewers have seen the film in presentations, classrooms, museums, and online. As the letters that flooded the Eames Office attest, the film has been particularly influential for scientists. As a work of trans-scalar constellation, it has received an audience wider than any other for its particular articulation of the universe and its vision of a scalar contiguity subtending it.

And yet, despite the powerful new medial technique employed, this was hardly a new model of the universe's scalar spectrum. The film's scalar shape morphed from *Rough Sketch*'s Einsteinian model to one that was . . . not Bohrian, but Newtonian. Newton had, after all, bridged the alleged difference between the expansive realm of the heavens and the narrow realm of the earth with his universal theory of gravity. As I discussed in chapter 1, for Newton the same law governs all scales, giving us a contiguous world to survey. This is no accident, as Charles Eames was a great admirer of Newton. In 1973 he and Ray had designed *Isaac Newton: Physics for a Moving Earth*, an exhibition for IBM that highlighted the life and achievements of the scientist. Despite Morrison's attempts to import his under-standing of quantum chance and rupture to the project, his contributions were easily subsumed within a filmic apparatus characterized by transcen-dental contiguity rather than immanent variation.

MORRISON'S CURIOSITY CABINET: KNOWLEDGE PRODUCTION
AND DISCIPLINARY FEEDBACK

The outcome of another series of scientific, artistic, and legal negotiations, *Powers of Ten: About the Relative Size of Things in the Universe* (hereafter "*POT* book" to differentiate it from the film whose title it shares) was pub-lished by Scientific American Library in 1982. Charles Eames had died in 1977, not long after *Powers of Ten* was released. Ray Eames collaborated on the *POT* book, but it was primarily the product of Phylis and Philip Mor-rison. While the book is centered on the scalar journey of the 1977 film and duplicates its artwork, the film is remediated in several significant ways. Each discrete scale is represented by a single image, reproduced from the film's artwork, on the right-hand page, supplemented by smaller images and explanatory captions on the left. These additional images, comprising photographs and other forms of historic media (textual passages, wood-cuts, etc.), are generally of objects found at the given scale. Each scalar shift thus becomes an ecological two-step: a milieu jump up or down of one power of ten (as the reader turns a page, revealing the hierarchically centered photograph from the film), punctuated by a lateral excursion to consider, on the facing page, objects that belong to the present scale but lie outside the central field of view produced on the right.

No smooth zoom is possible here. Time is frozen at each scale, with the viewer afforded the chance to peruse a mini catalog of representative ob-jects. As a result of such attention to scalar context, all but two scales in the book contain descriptions of intrascalar interactions or dynamics, narrowly beating out even *Cosmic View*.[55] Even with all of its rich contextual supple-mentation, however, the book produces fewer trans-scalar dynamics than does the 1977 film. Figure 20 shows a relatively decentralized interscalar structure, with local groupings or communities formed by mutually shared dynamics but no highly connected (privileged) nodes. Composed of mostly dynamic (active) nodes arranged nonlinearly (bidirectionally navigable) but with many path constraints and a relatively long average path length of 9.25 scales, the scalar dimensions of this work resemble *Cosmic View* (average path length, 10.75) more than *Powers of Ten* (7.35). Nonetheless, the distinctive "fish" shape of the graph (generated, it must be emphasized,

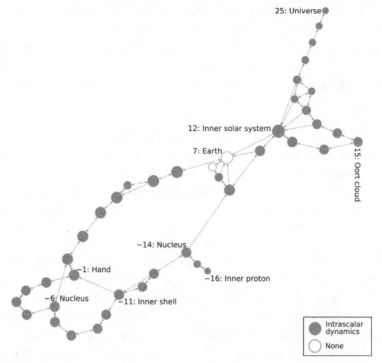

Figure 20 ▸ Network graph of all interscalar references in *Powers of Ten: About the Relative Size of Things in the Universe.*

by the same algorithm and parameters as every other scalar network graph in this book) reveals an additional set of scalar affordances: the "head" neighborhood contains a potent mix of small scales (from the human body to the atom and the cellular), while the "tail" neighborhood is composed mostly of interconnected galactic scales, with the inner solar system occupying a privileged position in the network as an obligatory passage point, a "gateway to the stars" that must be traversed if we are to connect the galactic to the realm of organic life, the tail to the head. To get from the tail to the head or vice versa, one can route either through the regional scales of the earth or the inner void of the nucleus; this network affords path redundancy in this vital "middle" section. Once again, the scalar structure of this work is highly singular and constructs certain implicit scalar potentials for virtual navigation and interaction. It is entirely appropriate, then, that it should be shaped like a fish or marine mammal: it is a virtual organism.[56]

What can this scale-mediating fish do?

Unlike all previous instantiations of the cosmic zoom, the 1982 book begins at the scale of the universe and progresses down the scalar spectrum, mimicking only the "inward journey" from the earlier Eames films. This change, "to begin with the largest dimension and descend to the smallest one," was suggested by the publisher, Howard Boyer, "so one can more readily discern the change in magnitude, as opposed to zooming out, zooming back, and then zooming downward."[57] Visually, then, the book attempts to foreclose the conceptual possibility of reversing or changing directions along the scalar spectrum. Consistency is key here, according to Boyer: a single motion is sufficient to experientially capture all scales. Because that motion is one of "descent," the viewer begins with the overview of the cosmos and eventually resolves the familiar in a continuous scalar skydive. From a design perspective, Boyer is surely wise to simplify, and neither Ray Eames nor the Morrisons objected. Yet the codex format, as I noted when discussing *Cosmic View* in chapter 2, is inherently bidirectional: any page can be turned in either direction, any motion reversed. By retaining the film's single, centered axis but pruning its scalar directionality, the *POT* book, it seems to me, actually encourages such reversals. The potential of the now fully reversible motion was so clear that in the 1990s, Eames Demetrios, the grandson of Charles Eames (from Charles's first marriage) and by then longtime director of the Eames Office, remediated the *POT* book into *Powers of Ten: A Flipbook*, which retains the monodirectionality of the 1982 book and removes nearly all contextual and narrative cues (most obviously the left-hand pages and paratextual matter), inviting the rapid flipping back and forth between descent and ascent.[58]

The bidirectionality of the 1982 book, however, is complicated by its left-hand pages, which, as I have noted, punctuate the axial motion with a lateral aggregation of similarly scaled objects and processes. The book acts on us in two directions: its gravitational force pulls us downward, while its disciplinary will swerves us sideways into a series of curiosity cabinets analogous to those of an amateur eighteenth or nineteenth-century naturalist: charmingly eccentric, subjective, and incomplete. For example, at 10^{-2} meters, opposite a square centimeter of the man's skin, are photos and descriptions of Japanese typewriter keys, buttons, a mushroom, a bombardier beetle, a hydromedusa (a type of tiny jellyfish), a fingertip (presumably attached to a finger), and a sample of foam held between two glass plates. Taken together, these facing pages almost form a separate book. Here, instead of the vertically oriented, controlled scalar descent of *Powers of Ten*, we receive a series of taxonomic slices from different disciplines. Astronomy gives us "A Garden of Galaxies" (10^{21}), urban planning gives us three versions of "Downtown" (10^4), chemistry gives us "Molecular Models" (10^{-9}), and so on. Philip Morrison's introductory essay, "Looking at the World," describes the panoply of scales as a function of disciplinary knowledge, noting the proper scales of numerous disciplines, including hydrology (approximately 10^6 meters), geology (10^7; "within this generation, geology has far extended its grasp"), physiology (10^0 down to the study of the cell, where microbiology takes over), molecular biology (10^{-7}), and so on. Morrison notes that at the extremes of scale, the number of available disciplines thins out:

> The two ends of our procession of images, the terminal scales of large size and small, mark the limits only of contemporary knowledge. On the one end, far out where the galaxies appear like a glowing froth in darkness, all our sciences become only one: cosmology. We know of no spatial novelties beyond the billion-light-year point.[59]

The scalar extremes are limited, not absolutely, but by detectable difference, seen through the lens of specialized disciplinary technique. We might call this "disciplinary resolution" and note that it is a function of technological development (in optics, sensors, spacecraft, computers), conceptual development (theoretical physics, cosmology, philosophy), and the development of disciplinary techniques of boundary policing, education, publishing outlets, social norms (as reified in professional conferences, publications, and funding streams), and scalar stabilization. Bruno Latour calls this process the production of "mobile immutables," practices and protocols of inscription that crystallize facts about objects into easily communicated and synthesized forms. These produce a plane of signification that unify a set of observations, practices, and phenomena into a discipline: "The manipulation of substances in gallipots and alambics becomes chemistry only when all the substances can be written in a homogeneous language where everything is simultaneously presented to the eye." This is a scalar process: scalar alterity has to be reduced to a conceptually planar form in which its transscalar objects can all be laid out side by side, and thus grasped together as a resolved whole. As Latour notes, "billions of galaxies are never bigger, when they are counted, than nanometer-sized chromosomes; international trade is never much bigger than mesons; scale models of oil refineries end up having the same dimensions as plastic models of atoms."[60] Each scalar domain is resolved separately, but it catalyzes a unified discipline only when a sufficient number of its entities enter the human scale as a tableau around which techniques and identities can stabilize. Such are the left-hand pages in the POT book. Each discipline arises only through the gaining of adherents, organized around such crystallized planes of inscription.

Morrison's particular formulation of the scalar structure of knowledge production in the natural sciences places great emphasis on its tight coupling with differential features of the universe itself. The production of knowledge at scalar extremes tapers off not only because we haven't invented the proper devices or disciplinary techniques to fill in these epistemic gaps but because we are running out of differences ("spatial novelties") to detect. The resolution of difference, made possible by conceptual, technical, and disciplinary innovation, also requires actual difference in the structure of matter and energy. Morrison thus describes a virtuous circle of knowledge production wherein both intrascalar and interscalar difference, when engaged through various scale-defined techniques, further bifurcate those techniques themselves:

> The swift motions of the air, its clouds and ceaseless winds, the slower flow of rivers, ocean currents, glacial ice, and the majestically slow drift of the solid continents themselves lie behind the single views. These occupy the dynamical sciences of meteorology, oceanography, hydrology, and geology.[61]

Singular scales, such as the continental-oceanic described here, contain an enormous degree of difference—such as differential rates of flow, degrees of solidity, phase states, and levels of internal variability—across many virtual axes generated by humans through engagement with the flows of matter and energy around them.

Disciplinary structures produce knowledge by stabilizing scales of inquiry, but the circuit is only completed when the world answers back, reveals its continuous differentiation, challenging its constituent parts to undergo further steps in this process. This is the disciplinary version of the circuit of scalar mediation that I diagrammed in chapter 1 (figure 1). Human knowledge production, as embedded in this flux, is internally differentiated as it goes about its business of detecting difference everywhere around and inside its own structures. The point here is that there are virtual and actual dimensions to this process of differentiation, and a productive feedback loop between them requires that neither be collapsed or foreclosed, as with many transcendent conceptual innovations that turn out to short-circuit their own productive processes: transcendent cosmologies that either place structural determinants outside the flux of the world itself (monotheistic religions, the immortalizing of the soul, Platonic forms) or deny this flux by positing a fixed universe incapable of further differentiation (determinism). Isaac Newton, as the thinker who most perfectly marries these two transcendental tendencies, stands at the summit of the contrary philosophy, the denial of scalar difference.

As I noted previously, Charles Eames was rather taken with Isaac Newton and created a traveling exhibition for IBM devoted to the scientist, *Isaac Newton: Physics for a Moving Earth*, in 1973. The exhibition was accompanied by a short film, *Newton's Method*, and followed up, in 1974, with a limited-edition set of Newton cards for IBM.[62] Philip Morrison, on the other hand, devotes multiple essays to excoriating Newton. How is it that Newton could prove so polarizing for these two scale thinkers and *Powers of Ten* collaborators?

At the beginning of book III of his most famous work, *Mathematical Principles of Natural Philosophy* (generally known as "The Principia"), Newton outlines his three "rules for reasoning in philosophy":

> **RULE I.** *We are to admit no more causes of natural things than such as are both true and sufficient to explain their appearances.*
>
> To this purpose the philosophers say that Nature does nothing in vain, and more is in vain when less will serve; for Nature is pleased with simplicity, and effects not the pomp of superfluous causes.
>
> **RULE II.** *Therefore to the same natural effects we must, as far as possible, assign the same causes.*
>
> As to respiration in a man and in a beast; the descent of stones in *Europe* and in *America*; the light of our culinary fire and of the sun; the reflection of light in the earth, and in the planets.
>
> **RULE III.** *The qualities of bodies, which admit neither intension nor remission of degrees, and which are found to belong to all bodies within the reach of our experiments, are to be esteemed the universal qualities of all bodies whatsoever.*[63]

Nature, claims Newton, "effects not the pomp of superfluous causes." Therefore, we are to assume, in our engagement with the world around us, that when we observe similar effects, they have the same underlying cause. Further, we can infer from similarly observed effects, and therefore from their assumed common cause, that all actions on all *unobserved* bodies also share the same underlying cause.

Newton's argument for universal causes is not derived from his mathematics. It stems, rather, from his theology. His two great theological enemies are the Catholic Church, which proclaimed the tripartite nature of God in the form of the Holy Trinity, and a Spinozan conception of God as contained within and suffusing the universe itself. Newton argues at length against the Trinity in his *Historical Account of Two Notable Corruptions of Scripture*.[64] Where the Greek text of the Bible discusses only "the spirit, water, and blood," Jerome changed the Latin texts to read "the three of heaven," leading over time to the establishment of the doctrine of a tripartite divinity (the Father, the Son, and the Holy Ghost). These must, argues Newton, be three separate entities, not one, for their unity would dilute the nature (essence) of God, making Him part of the world and part of heaven simultaneously.

In Newton's conception, God is unified, supreme, and has absolute dominion over the universe, as only one outside of and separate from his creation can possess. "This Being governs all things, not as the soul of the world, but as Lord over all."[65] Spinoza's conception of an immanent God thus represents a dilution of God's will and transcendence over his creation as odious to Newton as the doctrine of the Trinity: "a god without dominion, providence, and final causes, is nothing else but Fate and Nature."[66] God cannot be part of the world if he is to dominate it, for he is above dominating aspects of himself (which would imply a further division). "This most beautiful system of the sun, planets, and comets, could only proceed from the counsel and dominion of an intelligent and powerful Being" and "must be all subject to the domination of One."[67] God rules the world absolutely, according to this principle of unity that is his essence. He would not split himself into multiple natures or beings in order to directly enter the world; rather, he discloses the nature of his divinity and dominion through revealed scripture, in particular, biblical prophecies:

> He gave this and the Prophecies of the Old Testament, not to gratify men's curiosities by enabling them to foreknow things, but that after they were fulfilled they might be interpreted by the event, and his own Providence . . . be then manifested thereby to the world. For the event of things predicted many ages before, will then be a convincing argument that the world is governed by providence.[68]

This is Newton's refrain: the world is governed by providence, from a transcendent position above. It follows a template, a divine plan, that is revealed through two axes, one mechanical and one eschatological. The first is gravity, the second prophecy. Gravity affects all matter equally; it is trans-scalar. Objects near and far, cosmic and tiny, move according to a single universal force, the hand of God. As with mechanics, so too with prophecy: the dimension of historical becoming is predetermined by God's plan. The only convincing demonstration of this requires the plan to be foreknown, but in a code that can be cracked only through its application to past events. Scripture reveals God's plan retroactively, just as gravity reveals God's efficiency as final cause.

Consistently, then, Newton figures the production of knowledge as a process of decoding, the veracity of which can be determined on the basis of unity. For Newton, scale does not exist in any robust sense: objects of different sizes are animated by a consistent, singular force as the expression of God's will. The popular story, probably apocryphal, of Newton's apple captures this logic perfectly: One day Newton is sitting under an apple tree, contemplating the motions of the heavenly bodies, when an apple falls near him (or, in some tellings, strikes him on the head). Gravity's effect on the apple leads him to connect its motion to that of bodies at astronomical scales, and he has the Great Idea: movement and force work the same way at all scales! There is no scalar difference. For Newton, scale is banal. Chance does not exist; nothing can happen that isn't precoded into the motion of things. The single mystery, for Newton, is God's purpose in determining the universe's future in this way, a purpose that is revealed, in stages, through prophecy. Mathematics can break God's code by reducing the seeming variety of matter to an absolute and consistent—and thus abstract—set of laws (the laws of universal motion). Just as biblical interpretation of prophecy (of which Newton regarded himself a preeminent practitioner) strips away the seeming contingency of historical events to reveal the underlying necessity of providence, mathematics strips away apparent difference to reveal the underlying unity of divine cause. While the Eames Office's Newton exhibition focused on Newton's physics and on artifacts from his time,[69] a complete explanation of Newton's scalar articulation of the universe requires both the spatial scalar collapse of the scaleless laws of universal motion (expressed as mathematics) and the temporal scalar collapse of divine providence (expressed as the interpretation of prophecy).

Morrison takes Newton to task again and again in his writings, but perhaps most directly in "The Wonder of Time," his essay on chance, becoming, and scientific knowledge. Here he extols the virtues of quantum mechanics, which finally toppled Newton's conception of the scale-free unity of all matter. Where Newton's model leads us to perceive the world as a snapshot fully studiable and subsumable as human knowledge, quantum mechanics reveals the flux behind any such facile freeze-frame, the complexity that belies the reductive model of universal gravitation, and the mutability of the supposedly celestial spheres:

> The solar system, where Newton's rather simple world model dwells pure, nicely accommodates a snapshot, taken at a distance, of all around us. But such passive observation will not survive dinner time! Stronger interactions prevail on earth; being and becoming are incessant, everyday imperatives. It is true that the stability of this world requires a physics beyond Newton. Our solar system follows Newton through two centuries of almanacs of precision because it is so isolated in the depths of space. Collisions with other stars would destroy orbits utterly.[70]

According to Morrison, Newton's model of motion and the universe, far from establishing scale-invariance, is deeply scale-biased: it accurately describes only one domain of scale and exports its properties to *all* conceivable scales. Morrison even implies that there is nothing necessary about the applicability of Newton's laws to even the scale of planetary motion: Newton's observations, far from revealing a divine plan, are merely the contingent result of having a lot of space between planets. In other words, Newton's cherry-picked example of planetary motion obeys his mathematical model not because it is an immutable physical law but simply because there are so few planets and no other consequential bodies resolvable at that scale, and thus very few novel events that take place. When little can happen, it is easy to fabricate "laws" governing behavior.

Newton's thought, for Morrison, is plagued by a need for purity of explanatory cause and the collapsing of scalar difference. His clockwork universe does not allow for novelty, for surprise, for becoming. In short, it does not allow for differentiation. As such, it is obviously and demonstrably a false picture of reality: "Life is not at all like solar eclipses; novelty and surprise, building a tangled complexity of events, are much more the essence of the world than is the serene dance of the planets, however intricately they weave." Quantum mechanics reintroduces contingency and chance into science and thus brings physics "nearer maturity."[71] Modern physics' grappling with chance takes the form of embracing scalar difference. Subatomic entities are nondeterministic, atomic and molecular entities possess universal structure, and larger entities inherit these properties to different degrees according to their scale. All properties and behaviors, contra Newton, are functions of scale: "Stars shine, planets are round, bridges remain geologically rather small, cells divide rapidly, atoms randomly vibrate, electrons disobey Newton, all because of scale."[72]

Morrison's introductory essay to the *Powers of Ten* book lists many examples of scaling asymmetry. The real world is not Euclidean—its shapes change function as they change size. Animals' nutritional needs and morphological features necessarily change as they get bigger, small objects are stronger than larger ones, and so on. "These facts imply that form follows not function alone but size, especially over large changes of scale."[73] Scale is not a representational tool to reveal unity, but rather a fundamental determinant, a kind of ontological differentiator.

What, then, are the implications for representing scale in the *Powers of Ten* project? Because knowledge production is embedded in a virtuous circle of material engagement and iterative disciplinary differentiation along scalar lines, representing scale as a smooth contiguity risks negating its defining differences as well as its essentially temporal nature: scalar mediation tunes to frequencies of uncertainty and emergence, resonates around the event. Here Morrison must acknowledge the complex relationship between the 1982 book and the 1977 film. The film's inexorable movement left no space for the exploration of difference that produces this knowledge circuit. "The original trip was a long and uninterrupted straight line: the changing view along the line was presented without a sidewise glance," note Philip and Phylis Morrison in their brief preface to the volume (which appears before Philip's essay on scale).[74] In the book, as I have argued, lateral excursions break up each hierarchical scalar jump and produce "exhibits" loosely organized according to disciplinary logic. However, the book, as a static set of pages, also forfeits the film's capacity to represent movement, so central to Morrison's conception of scale:

> The limitation of the static image is not simply that it lacks the flow that marks our visual perception of motion: Real change in the universe is often too slow or too fast for any responses of the visual system. The deeper lack is one of content. A single take belies the manifold event.[75]

A robust model of scale need not cater to the perception of motion at the temporal scale of the human observer—in other words, the temporal scale of the cosmic zoom. A scalar model should capture change, motion, the event—not necessarily calibrated to the human perception of movement, but to the spatial and temporal range appropriate to the entities in question. Representing the scalar other, then, requires an apprehension beyond the visual: "It takes more knowledge to convey change. While our attention is seized by the moving image, we usually forget that it is nothing but a subtle illusion induced in eye and brain by a multiplicity of still images."[76] Morrison's medial argument consists of acknowledging that while a book seemingly fails to capture the fluid motion of a film, a film too is only a series of static images. Ultimately, then, a book is no worse off when it comes to the capacity to trace change. The mediating mechanism for the representation of change is not the illusion of motion but the succession of iterative cycles of knowledge production and material engagement, what philosopher of science Ian Hacking renders as "representing and intervening," two sides of the scientific method and the backbone of his own unique brand of realism, which recognizes different ontological commitments implied by these two stages of the methodological circuit. First we create a model to understand some aspect of the universe, then we utilize that model to empirically test and discover new features. We represent in order to intervene, and we intervene in the light of representations.[77]

Observation, as Morrison reminds us, is also an intervention. The linear relationship between the size of planets' orbits and their orbital periods was only discovered by astronomers who plied the tools of their trade over long periods of time: "Without watching over time, we could not have found this rule."[78] This is a disciplinary activity, defined by protocols and constants designed to register change in certain systems at certain scales. Through iterative conceptual and material engagement, we discover rules behind phenomena. Some of these rules describe fundamental order, while others describe fundamental disorder, or chance. Scale, then, cannot be directly visualized without falling into the visualist's trap of too "simple a trust in schoolbook geometry."[79] When it comes to visualizing scale, we are visualizing knowledge production itself. The *Powers of Ten* book is a schematic atlas of our current state of scientific knowledge as a series of dynamic processes. These processes of knowledge production in turn comprise half of a circuit of engagement with the universe, ideally articulating their practitioners to becoming itself by ordering it on science's disciplined grid. These dynamics at radically different scales then produce further differentiation through the most conscientious participants in the circuit, reorganizing the grid of knowledge. Disciplines stabilize particular scales, which tune our senses, conceptual apparatuses, and instruments—in short, our processes of mediation—to narrow slices of the scalar spectrum, resolving new forms of difference that must then be fed back into our disciplinary frames.

This is not to say, of course, that this is how knowledge production actually proceeds in all or even most cases. As Latour notes, beginning in the nineteenth century, "every object was thus divided in two, scientists and engineers taking the largest part—efficacy, causality, material connections—and leaving the crumbs to the specialists of 'the social' or 'the human' dimension."[80] These disciplinary asymmetries of legibility between different forms of causal explanation can lead not only to absurd disciplinary disputes but, more destructively, to extreme forms of reductionism. Thus, my interpretation of Morrison's model of disciplinary resolution in the *POT* book should be understood as an ideal model, not an empirical one. It calls for fluidity of disciplinary differentiation, a willingness both to divide the world into scales and to respond openly to the differential detail and explosion of new entities that such resolving cuts reveal.

On this reading, the *POT* book also serves as a partial corrective to the paradoxically scale-free Eames films that it remediates. The first film is Einsteinan and the second Newtonian in its scalar aesthetics. The Morrisons, in reusing *Powers of Ten*'s original image plates but disarticulating them from the bifurcated temporality, arcs of resolution, and singular subjectivity of the Eames films, actually reverse their meaning.[81] In the context of Philip Morrison's theories of scale and distaste for the theology-physics of Newton, the specific but subtle context of this remediation is clear: the 1982 book aims to root out Charles Eames's Newtonian aesthetics.

AFTERSHOCKS: CLIMATE CHANGE AND KNOWLEDGE PRODUCTION

The philosophical and representational debate between Charles Eames and Philip Morrison about the proper aesthetics and epistemics of scale, performed so productively in *Powers of Ten* and the *POT* book, has not dissipated in our larger culture. In 2016 Scientific American released a new version of the *POT* book, *The Zoomable Universe* by Caleb Scharf. Its debt is more to the Morrison book it updates than to the Eames film, but the standard cosmic-zoom tensions remain, as the book struggles to find a balance between smooth contiguity and scalar difference. Its text emphasizes scalar singularity, pointing out the radically different processes that occur at particular scales—and the disparate mathematical equations needed to describe them. At 10^{-9}, it notes that "on these scales the universe is a place of probabilities, of statistics, a dance of a multitude of branching pathways and curious relationships. That weirdness is at the heart of reality—it is what lets us exist."[82] Similarly, like Boeke, Scharf problematizes the visual modality as an adequate representational practice beginning at 10^{-7} (the scale of a virus): "The virus doesn't so much appear directly. Instead, it merely muddles the light."[83] Nonetheless, and tellingly, the accompanying image of the virus is composed of sharp, digitally rendered lines and vibrant colors.

Indeed, the publishers of *The Zoomable Universe* chose to have all of its images created by a digital artist rather than using scientific imagery, which had been the hallmark of the *POT* book (following the film). The effect is to erase the disciplinary scalar optics that are central to Morrison's project: this new book is about what we know of other scales, not how scale itself organizes knowing. By utilizing digital artwork for all entities at every scale, Scientific American's new book also reintroduces something of the Eameses' smooth zoom aesthetics: everything looks curiously the same here—it's Photoshop all the way down.[84]

The Zoomable Universe does cluster ranges of scales on the scalar spectrum, providing multiple levels of resolution simultaneously: the scalar quadrat, the spectrum range, and the totality of the spectrum. For instance, 10^{4} through 10^{8} are grouped as "A World We Call Earth."[85] This clustering helps to emphasize trans-scalar processes—the earth may be resolved as a *globe* only at the largest of these scales, but the bends of a river in Kenya (shown at 10^4) remain conceptually resolved as flows of the same surface. But with the evacuation of Morrison's self-reflexive disciplinarity, the scalar ladder, however vibrant, conceals its politics. What, the book does not ask, are the *consequences* of resolving the world with one scalar frame rather than another?

Despite some oblique references to human impacts on the planet, *The Zoomable Universe* is conspicuously silent about the great scalar challenge of our time: global climate change. This is a challenge that requires a robust scalar epistemics, not a commanding overview. To address it, multiple disciplines, each with their own resolving cuts, must work together to articulate a trans-scalar constellation that could meaningfully respond to the climate crisis by inspiring or compelling change. A crisis like this cannot be fully resolved at any one particular scale: not the planetary (atmosphere and space), not the molecular (CO_2 molecules), not the household (consumption and recycling), not the industrial (production and burning of fossil fuels), not the national (energy policy and subsidies, borders and trade), and not that of the human brain (consumption and reproductive behavior, acceptance or rejection of trans-scalar concerns). Scales must be unified while remaining distinctly resolved rather than collapsed. The production of knowledge must be self-reflexively tied to scale. These are the politics of trans-scalar constellation.

The Intergovernmental Panel on Climate Change, in each of its reports, constructs a cosmic zoom with an embedded multidisciplinary network that constitutes its politics. This is one vital example of the cosmic zoom as knowledge practice that follows the basic template laid out by Morrison. Of course, as we are all too aware, the IPCC's reports have failed to catalyze change at the scales required to affect climate change. This is partly because they compete with other cosmic zooms with different network structures, different bifurcations of the scalar spectrum, and different privileged scales. Scale is an ideological battleground. However, the IPCC's failure to catalyze multiscalar change is also a function of its disciplinary inadequacy. For all of its disciplinary diversity, it cannot incorporate the modes of knowledge production at the core of the arts and (post)humanities.

Climate change is not merely a crisis on a large scale but a crisis of scale itself. As Gabrielle Hecht notes, "Scale is messy because it is both a category of analysis *and* a category of practice."[86] Until we can align the two, it is scale itself that "muddles the light." In this context, multidisciplinary scientific reports will not suffice to address the needs of the future. We need to understand how we understand scale, how scales are stabilized, and how different forms of knowledge production produce different trans-scalar constellations. This implies both that the humanities need to work harder to problematize their own scalar biases (a process I link to the unfolding of a robust posthumanities) and that we all need to work harder to articulate different modes of knowledge together in productive, trans-scalar ways. One implication is that the environmental humanities must move far beyond their role of *communicating* science (dressing up IPCC reports, as it were) and enter into more meaningful multidisciplinary articulations of the scalar spectrum. All disciplines are in the business of making resolving cuts; the posthumanities can perform the necessary scalar politics by problematizing resolution and stitching these cuts together into catalyzing constellations.

A key insight of the 1982 POT book—its underlying scale-mediating fish—is that disciplinarization must serve the larger circuit of scalar mediation rather than blocking it in misguided attempts to fix knowledge and its underlying protocols into rigid hierarchies or reductive totalities. Resolving cuts should never be taken as exhaustive explanatory or even descriptive representations. They should not be taken as representations at all, but rather negotiations between ontological scalar difference and disciplinary resolution. How can we keep our knowledge production supple enough for this task? An understanding of disciplinary scale, explored further in the following chapter, begins to do this work. As I have attempted to show, scalar mediation is a circuit animated by ideology at the same time that it discovers new entities at new scales. It is an enrollment process that is nonetheless always startled by its discoveries, and thus continually affords opportunities to recalibrate the relationship of the human to knowledge production. Scalar difference itself, even in the form of the curiosity cabinet, opens the door to a reanimation of thought.

scale and difference

Toward a New Ecology

DISCIPLINARY SCALE

The previous three chapters have been building, piece by piece, a theory of scalar difference. Kees Boeke's groundbreaking work, *Cosmic View*, not only inaugurates the modern cosmic-zoom project but provisions us with an understanding of scale as a negotiation of difference between two (or more) surfaces. Because any two surfaces, unless they are identical, differ in size, resolving power, and the typical entities and processes that unfold upon them, the primary relationship that any medial system establishes between them will be scalar; that is, detail will be transferred from one to the other, transformed in size and density along the way. Chapters 3 and 4 explored the questions of temporal scale and epistemic access inherent in all such scalar negotiations, using the films and books of Charles and Ray Eames and Philip and Phylis Morrison as case studies. Chapter 4 also introduced the concept of scalar constellation, the larger assemblage of discrete scales and their modeled interconnections produced by scale-mediating works.

Because this book is concerned with the cosmic zoom, which attempts to sample a large swath of the universe's scalar spectrum, the examples of scalar constellation I've examined have presented us with startling models of the cosmos. Among other methods, I've attempted to visualize these models as network diagrams, emphasizing the latent scalar politics that such diagrammatic approaches reveal. This chapter will attempt to bring these various insights and approaches together into a more fully elaborated interdisciplinary theory of scale. Its interests are far less historical than those of previous chapters, and it explores only a single work of the cosmic zoom: the obscure 1999 CD-ROM version of *Powers of Ten*. To begin, let's take a deep breath and discuss the relationship between scale, epistemology, and ontology. Or, to put the question more directly, "What is real about scale?"

As I discussed in the introduction, different disciplines have distinct techniques for recognizing, recording, and theorizing scalar difference. Scaling effects are particularly important and widely discussed in geography, mechanical engineering, and biology. Most of Philip Morrison's examples in the 1982 *Powers of Ten* book are taken from the latter two disciplines, which often frame as their objects of study entities with which we are experientially familiar: bridges, airplanes, ants, humans. Engineers have, of course, always been aware of scalar difference; when it is left unaccounted for, their creations break. No mechanical design can be scaled arbitrarily without its strength, elasticity, and other properties changing. Thus, with every attempt to extend engineering to new scales (as with skyscrapers and nanostructures) comes increased scrutiny of "size-scale effects" for various materials.[1] Similarly, biologists are acutely aware of the effects of scale on morphology. "Size," argues evolutionary biologist and ecologist John Tyler Bonner, "is a supreme regulator of all matters biological. . . . It is the supreme and universal determinant of what any organism can be and can do."[2] Assemblages at different scales simply cannot function in the same manner. The very act of scaling, then, whether at timescales measured in nanoseconds or in billions of years, necessitates and produces changes in structure and function. Bonner synthesizes five "size rules" for biologists:

Rule 1: Strength varies with size.
Rule 2: Surfaces that permit diffusion of oxygen, of food, and of heat in and out of the body, vary with size.
Rule 3: The division of labor (complexity) varies with size.
Rule 4: The rate of various living processes varies with size, such as metabolism, generation time, longevity, and the speed of locomotion.
Rule 5: The abundance or organisms in nature vary with their size.[3]

Together, these rules require a material and functional differentiation be-tween any two biological entities that occupy different sizes. A relationship of scale is one of absolute difference. For biologists and engineers, the laws of scale are both descriptive and prescriptive. That is, we can empirically register the scalar differentiation of nature (description), and scale, as much as materials and design, will act as a fundamental determinant for the func-tioning of any assemblage (prescription). As biologist D'Arcy Thompson puts it in his classic *On Growth and Form*, "The effect of scale depends not on a thing in itself, but in relation to its whole environment or milieu; it is in conformity with the thing's 'place in Nature', its field of action and re-action in the Universe."[4] These limitations, imposed by the complete "field of action" affecting the object of study, are internalized by scientists and stabilized as disciplinary practices.

Andrew Pickering, in his analysis of scientific knowledge-making as a "mangle of practice," suggests that a discipline arises as scientific practice becomes socialized and generalized, and thereafter serves to limit human agency. Scientists are free to generate any goals (research questions) they wish, but the process of "transcription," or investigating a phenomenon, is strictly limited by the protocols and rules established within a discipline: it "carries scientists along, where scientists become passive in the face of their training and established procedures." Discipline, on Pickering's account, accrues its own agency, which it asserts against human agency by "[leading] us through a series of manipulations within an established conceptual system."[5] The scientist is forced to think in certain ways and perform certain actions as she goes about gathering empirical evidence. She is limited by the conceptual structure of her discipline as well as by the protocols and capacities built into the instruments she relies upon to gather data.

Both conceptually and technically, scale tends to act within processes of disciplined knowledge production as a filter that isolates certain features of matter in a given inquiry while occluding others. Bonner's size rules, for instance, all assume that the object to be investigated is the typical form taken by a single species. The discipline of biology has stabilized these rules as a common procedural language that enables relatively frictionless modes of inquiry and exchange among many different scientists. The rules, which arise in the processes of investigating actual organisms, give us useful (descriptive) information about their tendencies but also circumscribe the sorts of objects that can appear through the biologist's conceptual lens. Thompson notes that size rules are dependent upon milieu, but here he is beginning to span scales, and thus disciplines (from biology to ecology). As Pickering notes, this disciplining of knowledge practices gives necessary direction to all experiments, and thus "makes possible the emergence of resistance in conceptual practice." That is, disciplinary protocols force scientists to go about their work in certain ways beyond their control, and thus "push the experiment along until it collides with something—where the resulting pieces don't fit together."[6] Scientists, compelled by their disciplines, are constantly running into brick walls when the material universe fails to respond according the expectations generated by their protocols of access. Scale, as Philip Morrison noted in chapter 4, defines many disciplines; it provides the disciplinary framework within which certain phenomena become legible as objects of inquiry. Discipline produces resolving cuts that isolate certain structures in the matter-energy manifold as individual objects. If size rules arise through disciplinary empirical investigation, then they are only made possible through prior, more fundamental, upstream scalar differentiation into disciplinary protocols. Scale is thus an integral player in the mangle of practice: it is both co-constitutive of disciplinary protocols of scalar access and a key attribute of the objects that emerge through disciplinary agency. This is one of the reasons that it is so difficult to get a handle on scale: we can only describe it using the already scalar protocols of our disciplined knowledge.

The empirical description of scale rules, and the prescriptions they entail, only emerge once objects have been disciplinarily stabilized. In the realm of physics, for example, practitioners describe absolute scalar requirements and constraints for the entities within their disciplinary domains but treat scale itself as a second-order taxonomic category—that is, as one attribute of an object among many others. Biologists do not classify organisms according to scale. Engineers begin with materials, not scales. Physicists begin with particles, waves, and forces (though these are not always strictly differentiated) and then ask about their interactions and effects at different scales. In all of these cases, the question of scale arises only after an object, actual or virtual, has been defined. For the disciplined knowledge producer, first there is an object, then it occupies or "possesses" a scale in relation to some reference unit. This is as it should be: to be disciplined is to restrict oneself to a slice of reality, to register only that which is rendered visible by a constrained set of filters. As Morrison might put it, the "manifold event" must be reduced to certain manageable axes of measurable variables if we are to discipline our knowledge.

And yet, if it is true, as Morrison suggests, that disciplines themselves stabilize around particular scales and, as I have suggested, that entities en-

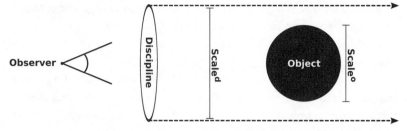

Figure 21 ▸ The disciplinary scalar lens. Scale^d resolves certain assemblages as objects, which can then be measured for relative size (scale^o). Whether we examine scale^d or scale^o, scale itself remains occluded. Diagram by the author.

countered through disciplinary lenses only don scales as a second-order consideration, then scale, seemingly all-pervasive, has nonetheless been evaded on either side. For disciplined knowledge producers, operating under particular (explicit or tacit) scalar constraints for pragmatic reasons, scale is a filter rather than a primary category. Each discipline acts as a scalar lens for the resolving of certain objects. Figure 21 illustrates this dynamic, and reveals a circularity: the knowledge producer sees objects and can measure their scale (scale^o), but this determination of scale is only enabled by the earlier disciplinary resolving cut that delimited the scale of the inquiry (scale^d) and thereby resolved particular assemblages or parts of assemblages as individuated objects.

Though produced by a different mechanism, the result is similar to what Harry Collins calls "the experimenters' regress," or dependence on experimental replication to confirm or disconfirm a researcher's interpretation of experimental (empirical) results. That is to say, scientists invoke disciplinary rules to qualify results as admissible knowledge or not, but in cases of interpretative dispute there can be no authoritative determination regarding whether or not the rules have been followed adequately. Such a determination requires comparison between the experimental outcome obtained and the experimental outcome that should have been obtained, and that is precisely what is under dispute in cases of newly discovered phenomena. If no clear criterion is available as to what counts as a correct outcome, "the experimenters' regress can only be avoided by finding some other means of defining the quality of an experiment; a criterion must be found which is independent of the output of the experiment itself."[7] In other words, interpretive disputes have to be solved at scale^d, not scale^o. The experimenters' regress obliterates any hope we have of resolving scientific disputes with further experimentation. Instead, social factors step in to resolve the crisis, adjudicating the dispute and achieving consensus, at which point the discipline's lens has been slightly changed or enlarged, or alternatively, has resisted admission of any new truths.

As Collins notes, the period of interpretive uncertainty does not last long, and once a consensus judgment has been made, it provides a criterion by which empirical practice may be evaluated, relegating results that don't fit to the "incompetence" or "pseudo-science" file. This cleanup process provides retroactive assurance that there was, all along, a proper way to obtain the necessary result. Collins calls this process "crystallization" and reminds us that "even for [scientists] it is very hard to recapture the uncertainty of the time of creation once the debate is closed and the correct way of going on has been crystallized into the new scientific institutions."[8]

In much the same way, disciplinary scalar optics crystallize the empirical fields in which objects may be resolved. Every object studied is determined to possess a scale, and yet it was the disciplinary scale that resolved it—and only entities at its scale—as a legible object in the first place. Thus far, this circularity is of the same variety analyzed by Collins and most science studies scholars: officially knowledge is certified as the outcome of disciplinary rules, but behind the scenes it is adjudicated by social factors that fix interpretations and thereby halt the regress. This is not to say that disciplines are absolutely constrained to the resolution of only particular scales. Most disciplines resolve a range of scales. My argument is that that range has been solidified or crystallized through disciplinary negotiation in order to contain chaos and preserve disciplinary boundaries: to order knowledge. New objects at new scales provoke boundary disputes until the range is recrystallized.[9]

Because the recrystallization process tends to erase the past uncertainty and open-ended negotiation, the scalar optics of a given discipline quickly revert to a relatively rigid framework for making resolving cuts. To uncover the disciplinary magic trick, we need metaknowledge practitioners, that is, scholars working within a different discipline with a different set of scalar optics that enable them to resolve knowledge-making in the process. Such metaknowledge workers can readily discern the parameters and protocols that make up a discipline's scalar lens; to them, the scale of the discipline (scale^d) is visible. Morrison's discussions and identifications of disciplinary scale in the POT book amount to exactly this. But even here, we are working in a circle, taking the objects of inquiry within a given discipline as evidence of the scalar assumptions that have already been enacted to stabilize that discipline. Whether we train our gaze on particular objects through the lens of one discipline, or on discipline itself (also through a disciplinary lens, of metadiscipline), the determination of scale arises only *after* an object or discipline has been fixed. The set of objects or disciplines is taken as a given, and scale arises as a description of certain size relations within this set.

In the past three chapters, however, we have again and again detected another level of scale, one that does not arise only as a measurement of size relations within a set of already given entities, but rather seems to precede the stabilization of entities as such, and indeed participates in that stabilization. What we are sensing, if I may presume that the reader has shared this intuition, is that scale functions as a dynamic framework that *composes* entities (as assemblages)—that it acts as a determinant of becoming. The reason that in biology, engineering, and physics scale seems to be prescriptive as well as descriptive is that scale itself is a compositional force that distributes matter-energy across size domains as a kind of scalar articulation: the assemblages produced function at multiple scales simultaneously, as interdependent multiscalar entities rather than as autonomous objects nested within one another. As I discussed in chapter 2, this is an aspect of what Karen Barad calls "mattering." It is also a characteristic of what Jane Bennett describes as "thing-power," or "a liveliness intrinsic to the materiality of the thing formerly known as an object"; of Félix Guattari's "machinic heterogenesis"; and of what Jay McDaniel described in 1982 as the "creativity and sentience" of matter.[10]

These thinkers and many others, from ancient Chinese and Greek philosophers to Lucretius's famous *clinamen*, or swerve of atoms that produce new forms,[11] have explored this ontogenetic capacity of matter. What has perhaps been overlooked, however, is the ineluctably *scalar* quality of matter's creative differentiation. We have been exploring the edges of ontological scale up to this point; now we take the plunge into the heart of scalar alterity itself. Scale, I argue, is a primary form of difference, a diagrammatic force of composition that continually differentiates itself from within, producing new objects of incommensurate sizes.

We are now in a position to understand why this ontogenetic function of scale has remained more or less unexplored. We can formulate it as a problem of metascalar knowledge. For disciplined knowledge workers, scale appears only after the fact as a set of relations among objects resolved through the scalar lens of their discipline. This is true whether scholars take nonhuman objects to be inert or vibrant. For metaknowledge workers, on the other hand, the scalar protocols of disciplines are visible as objects of knowledge. In neither case, however, does *scale itself* become visible. Scale, as compositional force, as prior to the entities it differentiates, does not, in our disciplinary form of knowledge production, ever *itself* become the object of knowledge.

DIFFERENCE AND IDENTITY

Is it even possible to speak about scale itself? Kees Boeke's attempt to impart a resolution-aware cosmic view to his students and readers, and Philip Morrison's fascination with the primacy of scale in the composition of the universe, for which he had only disciplinary and metadisciplinary language available, both point toward the need to theorize and discuss scale itself. But where can we find a solid base, or a solid vocabulary, to speak of scale in this way?

Italo Calvino, one of the twentieth century's great scalar storytellers, satirizes precisely this problem by way of a comical thought experiment in his story "All at One Point." This singular tale is an account of life before the Big Bang; that is, life before time and space. Here, everyone and everything are crowded together at one point:

> Just with the people I've already named we would have been overcrowded; but you have to add all the stuff we had to keep piled up in there: all the material that was to serve afterwards to form the universe, now dismantled and concentrated in such a way that you weren't able to tell what was later to become part of astronomy (like the nebula of Andromeda) from what was assigned to geography (the Vosges, for example) or chemistry (like certain beryllium isotopes). And on top of that we were always bumping against the Z'zu family's household goods.[12]

Of course, a punctiform universe is without scale, and thus without discipline: it is unstudiable, unknowable. Calvino's story functions as a narrative only through the conceit that the punctiform universe-to-be possesses the *potential* for an infinite set of differentiations, each of which implies a scale and a discipline to observe it. Calvino presents us with the following narrative dilemma: which comes first—objects, disciplines, or form? In other words, what happens at the moment of the Big Bang, which looms just beyond the threshold of this narrative? We might call the suspended animation of a punctiform universe a disciplinary singularity, but certainly some potential of differentiation exists even there. This remainder, what is left when we compress the universe to a single point, where no discipline is possible, whatever it may consist of, is scale itself. How can we unpack it?

We have a discipline for every scale but none designed to study scale itself. From a disciplinary perspective, is there is proper scale for scale?

Something akin to this problem is addressed by Gilles Deleuze in *Difference and Repetition*. Deleuze is not concerned with scale as such, but his treatment of difference provides us with the final element of critical scaffolding necessary to explicitly formulate the notion of scalar difference with which we have been flirting thus far. Just as we have been searching for a concept of scalar difference that would not collapse into either flat spatial contiguity or scale-invariant simultaneity, Deleuze was searching for a concept of difference that doesn't collapse into identity (the Same) or negation. Western culture, according to Deleuze, has always been so heavily invested in identity—as autonomy and self-sufficiency in the subject, as stable and self-identical entity in the object—that it has only been able to conceive of difference as secondary, even as monstrous. The different is that which isn't the Same, that which isn't X. First comes the object, then come its others. First comes our concept of X, then comes a distribution of differences derived from that identity. For Deleuze, this is backward. Here he follows the Nietzschean insight that multiplicity and force (Nietzsche's "will") are fundamental, with identity coming to be stabilized only afterward. This process of stabilizing identity proceeds by rewriting the original multiplicity into a one (a self, who does the willing), and thus obscuring the forces that produced it. In contrast, an adequate ontology must start with difference itself. For Deleuze, carrying forth this charge means nothing less than attempting a revolution in thought, an overthrowing of the supremacy of identity in our hierarchized conceptual economy:

> That identity not be first, that it exist as a principle but as a second principle, as a principle *become*; that it revolve around the Different: such would be the nature of a Copernican revolution which opens up the possibility of difference having its own concept, rather than being maintained under the domination of a concept in general already understood as identical.[13]

To reverse this conceptual priority, to start with difference and thence derive identity, is also to alter the role played in the process of thought by concepts in general. In the anthropocentric paradigm of identity, concepts primarily play a classificatory role: they define ideal types and thus serve as models by which to measure the deviation of encountered singularities. Difference is thus relegated to either generic difference between concepts or latitudinal difference (as deviation) between a singularity and the concept that it represents. Ontology thus becomes representational. This is the basic Platonic formula, which conceives of difference only in relation to representation. Identity is predistributed as conceptual, and difference is then distributed as a series of representations. Just as we could not forever deny the universe its relativity and dynamism while we followed Ptolemy in subordinating all of its movements to a fixed center (the earth), we cannot continue to deny the universe its multiplicity and becoming by following Plato in relegating ontological difference to representation: "Difference is not and cannot be thought in itself, so long as it is subject to the requirements of representation."[14] These requirements of representation make thought a copy, an analog, of something else, and thus contain thought within an axis of relative fidelity—that is, thought becomes subordinated to authenticating protocols in the form of questions such as "how accurately does this trace the contours of the thing it represents?" What is at stake is a freeing of thought itself from a merely representational dynamic to one that is animated by difference, that participates in the ontological movement of differentiation that suffuses all of matter-energy.

Differentiation, for Deleuze, arises out of intensity, understood as difference of degree within the univocality of being. Intensity is thus always internal to substance or being, which is to say, it is that which differentiates and divides within a type rather than that which opposes it from without. "Every intensity is differential, by itself a difference."[15] This internal difference, which requires no outside or other in order to differentiate itself, is the fundamental ground or starting point for all becoming. It nonetheless finds itself incessantly exteriorized and extended into differences of kind:

> Intensity is the uncancellable in difference of quantity, but this difference of quantity is cancelled by extension, extension being precisely the process by which intensive difference is turned inside out and distributed in such a way as to be dispelled, compensated, equalised and suppressed in the extensity which it creates.[16]

"Extensity" is difference explicated. This is the process by which difference itself (a relationship of different to different) is "drawn outside itself" and thereby subordinated to identity. This process of explication dispels the internal inequality that constituted the primary difference and reorders it as an exterior relationship between two things. Extension, in this sense, is representation, as one object is conceptually made to resemble or not resemble another object. The primal difference, "indistinguishable from depth in the form of a non-extensive and non-qualified *spatium*, the matrix of the unequal and the different," is thus concealed beneath the sensible to which it gives rise.[17]

Deleuze's understanding of difference is a radical revision of the common understanding of the term, but it is no less divergent from other common poststructuralist readings of difference that arose in the 1960s and continue to hold sway. To take perhaps the most influential, Jacques Derrida's notion of *différance* combines the mechanism of spatial displacement or differentiation with that of temporal deferral. Difference, on this account, is that which defers the moment of encountering the thing itself in an infinite play of substitutions. All attempts to grasp the different, the thing itself (any thing), leave us only with a sign pointing to something else. "Signs represent the present in its absence; they take the place of the present."[18] Derrida thus follows Ferdinand de Saussure and Jacques Lacan in characterizing language as a play of substitutions. His radical move, however, is to deny that there is a thing in itself at the end of the chain. Signifying substitutions produce endless difference (in the form of signs) by deferring a presence (being) that is perpetually absent. Thus, nothing can be said to ground or originate the play itself. Derrida originally formulated this as a freeing of the potentials of signification from the transcendental order that (falsely) guarantees a link between signifier and referent, enabling "the joyous affirmation of the freeplay of the world and of the innocence of becoming, the affirmation of a world of signs without fault, without truth, without origin, offered to an active interpretation."[19] Explicitly invoking Nietzsche, Derrida articulates freeplay to the Dionysian project of producing new values without recourse to or need for the authenticity conferred by historical origins. This form of creative affirmation is nevertheless purchased at a price too dear for Nietzsche: the banishment of ontology, or any attempt to refer to or encounter being outside of language.[20]

In any case, it was clear by *Of Grammatology*, published a year before *Difference and Representation*, that Derrida's method of reading tended not toward affirmation but toward deconstruction. This celebrated method begins by diagnosing, within a given text, one or more binary formulations that establish a hierarchical relationship among its terms. A careful reading of the play of signification then reveals, inevitably, that the dominant term relies upon the subordinate one in order to signify at all. In the end, we are left with the renewed realization that the text has produced its meaning "out of thin air," as it were, without any grounding authorization to do so. A powerful machine for the dismantling of linguistic hierarchies, Derridean deconstruction is thus an example of what Deleuze calls extensity or explication. Instead of reifying object identities by transferring them to the linguistic realm, Derrida deconstructs them and renders them equivalent through this transference. Ultimately, then, Derrida's understanding of *différance* as a linguistic operation has the effect of dismantling both secondary difference (relations between hierarchized entities) and primary difference (intensity). Without passing any judgment on the usefulness of the knowledge that Derrida produces, it is clear that his notion of difference is inadequate to account for or make use of primary differentiation, which takes place in the world of quarks, atoms, molecules, and assemblages, and is not confined to the world of textuality and representation. Paradoxically, Derridean *différance* reduces difference to a kind of heat death of linguistic flatness, a zero-intensity topography that elides both the primary differentiation of matter-energy and the dimensional horizon of the virtual that affirms it.

Our exploration of the cosmic zoom thus far has revealed the basic requirements for any theory of scalar difference: it must acknowledge that no two scales can be resolved simultaneously, and thus account for the mediation of scale as a negotiation between surfaces (scale as a material mediation); recognize that scales are stabilized by the mediations performed by and for particular entities (scale as experiential); account for the elisions of scalar difference that result from representations of scale as a contiguity of size domains or homologous processes occurring at different scales simultaneously (time and space possess unique interrelations at different scales that cannot be collapsed without eliding difference); and recognize that scalar difference precedes and exceeds the descriptive differences (genera) and prescriptive differences (laws) produced by particular knowledge domains (disciplines). Scale holds as a primary differentiation between *any* two assemblages, regardless of the particular stabilized scales they are represented as occupying.

I contend, then, that to satisfy these requirements, we must have recourse to a theory of difference that is itself capable of linking the primary differentiation of matter-energy to the production of stabilized domains of knowledge. That is, it must take difference itself as primary and prior to disciplinary empiricism, but also recognize the material feedback loops that occur when scales are stabilized and thus potentials of access and engagement are produced. Deleuze's theory of difference is my starting point because it is supple enough to do exactly this. A radical form of realism, it takes matter and energy as primary, denies any transcendental realm outside of the real—thus materializing thought and reining in metaphysics—and yet also retains for thought a *relatively* transcendental horizon in the form of the virtual that renders it a midwife of material actualization. This framework links matter-energy to thought, or materiality to culture. The Deleuzean theory of difference is thus uniquely capable of accounting for the paradoxes of scale that the cosmic zoom reveals.

PRIMARY SCALAR DIFFERENCE: BEFORE THE HUMAN

Scale precedes and exceeds the human, and yet scales are produced through processes of mediation. This is because difference is primary—it is a process of differentiation that occurs as the fundamental becoming at the heart of matter and energy itself. It is the result of intensity; it comes from within. Matter pulses, moves, differentiates itself into quarks, atoms, molecules, and so on. The active agent is matter-energy, not these stabilized (differentiated) entities. What I wish to argue, then, is that every intensity, every difference within matter-energy, produces a scalar relationship. This is to say that intensity, an internal relationship of quantitative difference, produces the differential of scale. This is primary scalar difference, which we can now locate in its proper place, at the birth of becoming. As one thing becomes something different, as it others itself from within, it is composed into a new assemblage, either by gaining or shedding articulations. It becomes larger or smaller, composed of a slightly different set of forces. Because all qualitative change depends upon these processes of intensification, of a one becoming multiple, a quantitative swerve must take place in all acts of becoming. This swerve is, and can only be, a shift in scale. In a crucial sense, then, scale *is* difference. Difference is scale.

We commonly conceive of scale only as a relationship between two given objects, but this is because we are trying to read scale backward, after we have stabilized particular scales and the objects that occupy them through our medial practices. Attempting to analyze scalar relations at this point, through a disciplinary lens, tends to relegate scale to a comparison of size. All of this requires units of measurement, reference scales, and epistemologically isolated entities to measure. The objects collected within the 1982 *Powers of Ten* book are an example of this size-based menagerie. This is secondary scalar difference, as measured between objects that have been, from a particular point of view and within a particular discipline of knowledge production, assigned stable (signifying) identities.

The problem is now clear: the Eameses' films and book subordinate scalar difference to identity. This is our common understanding of scale and what authorizes the smooth zoom of the films: scale as a measurable relationship between entities. Both ontologically and epistemologically, the objects come first. The cosmic zoom simply strings them together, purporting to reveal the spatial contiguity that underlies their experiential (perspectival) alterity. There is truth to this contiguity, of course, but it has the effect of reinforcing the scopic dimension of representation, containing scalar difference at the level of perceived size difference, where representation allows us to capture that which is experientially overwhelming. It is precisely this process that Kant refers to as the "mathematical sublime": immense magnitude initially overwhelms our senses, until our capacity for progressive calculation comes to our rescue, generating a "pleasure, aroused by the fact that this very judgment, namely, that even the greatest power of sensibility is inadequate, is [itself] in harmony with rational ideas, insofar as striving toward them is still a law for us." In such a case, it is our capacity to represent mathematically that reasserts mastery over scalar alterity. The pleasure highlighted by Kant, a kind of narcissistic rationalism produced when "the subject's own inability uncovers in him the conscious-ness of an unlimited ability which is also his," is enabled by the reassertion of identity over difference, a rearticulation of the universe's scalar hetero-geneity to a single, mathematically unified plane.[21]

Biologists and engineers similarly subordinate scale to size, and thus to identity. The difference between organisms or structural components is far more fundamental than that which can be expressed by a calculated or implicit ratio. Morrison digs deeper, pointing toward the fundamental dif-ferences in interactions at different scales, but he can still only perform this gesture through an empirical, disciplinary lens, by describing the entities and processes that his own disciplinary methods make available to him. Our debt to his insight and eloquence acknowledged, we must go further. This requires us to view scalar difference as primary, as a fundamental dif-ferentiation that is internal to matter and thus continually produces further elaboration, further scalar differentials, as it innovates entities. Each new entity is also a changed entity: it bears a relation to itself as a differentia-tion, in space as well as in time. This movement is always across scales, not because agglomerations of matter cross predefined thresholds of size but because material elaboration always produces new relations in the scalar spectrum: between the old and the new, between the assemblage formed between them and the outside, between the rhythms before and after the rupture of the new. Scale, as the differential of becoming, is the painful process of what Karen Barad refers to with great economy as "mattering."

I argued in chapter 1 that Barad's notion of "agential realism" enables us to conceive of mediation at the level of matter itself. Matter differentiates itself in a continual medial negotiation. We can now render explicit the primary relationship between mediation and scale: scale is an intensive process of differentiation (primary scale) that produces mediating agents (interscalar entities or primitive subjects) that inhabit and negotiate the space that continually opens between differentiating elements (secondary scale). This two-step dance continues *ad infinitum*, extending up through the great chain of being, now decidedly less hierarchical (the great chain-link of being, a mesh). There is thus a "push" and a "pull" to scale: on one hand, intensive differentiation propels matter-energy into new assemblages shifting up or down the scalar spectrum; on the other, the space that opens between entities has its own positive horizon, that of the virtual, which takes the form of negotiation between differential entities, now viewed as surfaces that resolve each other. This gives rise to the stabilization of particular scales for particular entities. "Ontology is the dice throw," claims Deleuze, channeling Nietzsche, "the chaosmos from which the cosmos emerges."[22] Scale is both the chaosmos and the cosmos, primary differentiation itself and the medial stabilization of a world.

SECONDARY SCALAR DIFFERENCE: NAVIGATING MILIEUS

The cosmic zoom takes as its subject secondary scale, which postdates and derives from a primary scalar differentiation. We might call this second-order mediation—the production of intertwined systems of knowledge, technics, sociality, and semiotics aimed at providing perspectival access to disparate scales—the production of a milieu. A milieu is an environment as it signifies for a particular entity or group of entities. A milieu is thus a perspectival field, a set of signifying elements or differentials unified by some form of access for a particular subject. When we colloquially refer to scales, we are referring to milieus, or scales stabilized as environments-for. As such, there are potentially as many milieus as there are perceiving entities—and vice versa, as a milieu reciprocally stabilizes its subjects.

The cosmic zoom approaches, again and again, from many angles, the paradox of scale: scales are arbitrary and constructed, but scalar difference is real. We can now parse this paradox as a productive synthesis rather than an antinomy. Once we come to view scalar difference as primary differentiation (not subordinated to identity), we approach the perspectival dimension of scale as a secondary difference. The primary process of differentiation gives rise to perspectival capacities—capacities to sense an environment, to sense difference itself, to stabilize objects; in short, awareness of difference or "mattering." The scalar negotiation that follows produces a milieu. Milieus can (and must) be shared, though they are different for different entities. The point here is that milieus are not fully arbitrary. They arise out of differentiation and interaction, and thus reflect material processes of transfer and exchange. A world is built, but its raw materials are matter and energy as differentiated quantities (primary scale). Nonetheless, the world that emerges in relatively stable form, perspectivally accessible to its own constituents as sensing entities, further stabilizes particular scales as defined size domains. By necessity these are arbitrary, but the difference contained within them is not. This is why the "powers" or scales in cosmic-zoom media are potent conceptual entities: they demarcate arbitrary domains, arranged in relation to typical human processes and techniques of access, but they capture fundamental dynamics of scale itself. Any two scales bear a derived and unsurprising mathematical relation to each other (mathematics is a technique of scalar access) but also dredge up a manifold of scalar difference, protean and irreducible, whether tamed by representation or not. Hence the agonizing problems faced by members of the Eames Office when they attempted to represent the entire human-accessible universe as a series of nested scales forming a single spatial-temporal contiguity. Scale, unlike abstract space or time, is by its nature discontinuous. Only a complex series of tradeoffs and occlusions, a counterrevolutionary suppression of scalar difference, makes the illusion of scalar contiguity possible.

Engineers face the same problem, as in 1940, when the designers of the Tacoma Narrows Bridge (then the third longest bridge in the world), who had assumed that "the flow pattern around geometrically similar bodies is geometrically similar, independent of scale," watched their bridge collapse into Puget Sound only four months after opening.[23] Scale can be occluded, ignored, or papered over, but it will bite back. Like any technique aimed at navigating an environment, the stabilization of (secondary) scale can get things wrong. This problematic, one of navigation, of fit, of usefulness, is primarily ecological. On one hand, ecosystems arise at various scales due to primary scalar differentiation. On the other, the milieus we build lead to success or failure, to creation or destruction, to subjugation or the affirmation of difference, partially on the basis of how they parse scalar difference—that is, how attentive they are to the interdependencies of material-energetic systems of which they are a part, and how responsive they are to the milieus of others, which exist at every possible scale. What kind of scalar access, we must ask critically, does a particular milieu afford? Which scales does it resolve, and how?

Postcognitive psychologist J. J. Gibson developed his famous theory of "affordances" to account for the robust interaction between organisms and environments initiated by mere perception. Prior to any cognitive map, he argues, all organisms directly perceive the movements and actions made possible, or afforded, by the environment around them. "The *affordances* of the environment are what it *offers* the animal, what it *provides* or *furnishes*, either for good or ill. . . . It implies the complementarity of the animal and the environment."[24] An environment, then, *just is* a collection of affordances. This is what we see when we see: potential interactions with surfaces. But surfaces do not come ready-made as a series of static images that we subsequently decode. Rather, surfaces, in order to be perceived, must be extracted as "invariants" from a continuous flow of sensation. Perception, then, already requires interaction with an environment: looking around, moving around, and so on. These kinetic processes take place within the mediums of air, water, and earth, which yield a flux of sensation, but within that continuous flow certain features do not change. These are surfaces, or the meeting points of mediums. Thus the sky is a surface, as is the lake, the prairie, the linoleum floor. Gibson's emphasis is on the primacy of surfaces and their textures (or "furniture") as potentials that are perceived as invariants within an interactively produced manifold. Thus concepts like "object" and "space" are later theorizations, not primary categories of experience:

> According to classical physics, the universe consists of bodies in space. We are tempted to assume, therefore, that we live in a physical world consisting of bodies in space and that what we *perceive* consists of objects in space. But this is very dubious. The terrestrial environment is better described in terms of a *medium*, substances, and the surfaces that separate them.[25]

The "ecological approach to visual perception," after which Gibson's final book is titled, poses ecology as an alternative model of interaction to the Newtonian understanding of the world. Bodies and "space" afford nothing and are thus not perceived features of the world but abstractions that elide and discretize the actual nature of movement within an environment, which is characterized by continual flow. If it were not, we would not be able to determine which interactions are possible and which aren't; this requires testing surfaces against a continually changing perspective to determine the position of their edges (such as the edge of a cliff) relative to the observer's movement: "invariance in a flow of stimulation."[26] To perceive this invariance within flux is to separate the perceiver's body from the features of the environment and thus to perceive what further movement would afford (a nasty fall, an elevated perspective, faster movement relative to a mobile food source, etc.). "To perceive is to be aware of the surfaces of the environment and of oneself in it. . . . The full awareness of surfaces includes their layout, their substances, their events, and their affordances."[27] The "meaning" of textures, as structured surfaces, is thus perceived directly. On the other hand, it is not possible to perceive "bodies" in "space" at all; it is only possible to *conceive* of these in a superfluous act of theorization.

For Gibson, then, perception is fundamentally an emergent awareness of the potentiality embedded in structured surfaces. As such, we must note that it is *resolution-constrained*. Texture, or surface structure, is perceived as a set of affordances, which are virtual relationships between the perceiving/moving assemblage and the invariant features of its sensory flux.

Textures with real affordances (which are always virtual until actualized) are those that can potentially interact in meaningful ways with the perceiving assemblage, which means that they have come to share a scale with that assemblage (what Gibson refers to as the body). Texture that is too fine or too coarse to contain affordances for a perceiver (too large or small to affect its bodily movements) cannot be resolved. Because of this affordance-resolution relativity, "for the terrestrial environment, there is no special proper unit in terms of which it can be analysed once and for all."[28] The environment is scale-relative.

Gibson's model of "ecological optics" clearly acknowledges and accounts for scalar difference. How does it bear on scalar *mediation*?

I have argued that Kees Boeke's book *Cosmic View* presents scalar difference as a drama of resolution driven by the periodic negotiation of difference between two surfaces: the medial surface of the page and the reference or primary surfaces of the Werkplaats Children's Community campus, the European continent, the earth, the galaxy, and beyond. As a narrative, the work resolves particular scalar relationships in a serial manner, suggesting to the reader that these stabilized dynamics form a single, aggregate surface upon which all of their differential features are present: radio waves, viruses, the girl and her cat, the Milky Way, and so on. Gibson makes no provision for this surface of difference itself, but his elaboration of the role of perception as a mediator between body and surface, enabled by the resolution of texture as affordance, lays the final paving stones for our theory of scalar mediation.

There is no need, in Gibson's model, to assume the prior stability and autonomy of a series of objects or subjects. It thus passes Nietzsche's test, which requires that we *account* for the construction of the subject out of primary forces rather than attributing action and will to an already constituted, singular subject: "In [the] future, concepts such as the 'mortal soul' and the 'soul as the multiplicity of the subject' and the 'soul as the social construct of drives and emotions' will claim their rightful place in science."[29] The differentiation of subjects and objects is what is to be *explained*, not the starting point for a confident description of objects outside of the (implicitly human) self. Perception, in Gibson's understanding, is itself a form of differentiation: out of the flux of energy and matter (waves and particles interacting, light reflected and absorbed, etc.) arise differentials in the form of relationships between movement and rest. These differentials grow as the result of an intensive magnification that "extracts" invariants from flows, which is to say, stabilizes difference itself into articulable assemblages. That which emerges as invariant is perceived as a surface with scale-dependent structure (texture), and the remainder is taken to be the body of a perceiver. This is itself a constant negotiation or resolution of difference. What Gibson calls an affordance is a virtual potentiality: it arises only out of differentiation-as-perception but gives rise to a qualitative relationship between an observer and a surface. It is thus part of an ongoing negotiation that stabilizes the subject as distinct from the surface it perceives—this is how subjects are made.

Rendered in the simultaneously medial and ecological language of surfaces, what I have described as "secondary scalar difference" emerges from this primary negotiation when a third surface is introduced as a medium.[30] This secondary or medial surface extends the negotiation of difference to a greater (potentially infinite) range of primary surfaces by stabilizing a series of relationships between its own dimensions and texture (resolvable detail) and that of its reference surface(s). This is scalar mediation. It is not, I stress again, representation. No surface represents some other surface; each surface is a structure at the meeting point of various mediums that contains its own affordances, not (as Gibson also stresses) a copy of something else. Each surface, as I have emphasized, must negotiate its difference with other surfaces; this process never collapses into mere representation. The map cannot be the territory because it possesses a different resolution and thus a different texture. This is a material affordance based upon a differential, and holds true even for cartographic surfaces produced at a 1:1 scale. Droplets of ink and the rocky ground possess different affordances and thus different embedded scales. Resolving their difference would mean stabilizing a scalar relationship between their respective textures, extracting potentials from this difference and thereby reducing them to concrete possibilities. This is the process of actualization, which requires the stabilization of a particular scalar relationship. The textures of the two surfaces enter into correspondence—a scale—as their resolutions are fixed in relation to the mediating agent, the observer for whom the scale is stabilized.

This is what authorizes Boeke to see the negotiation of difference between surfaces as a political project: it constitutes a milieu for its perceiver (reader) and differentiates matter and energy into specific affordances for action. This is a reductive process of resolution, a reduction of flux into stabilized relationships, an actualization. As a process of milieu building, the differentiation between surfaces and perceivers (and of course one perceiver is another's texture) constitutes our fundamental interaction with an environment. Ecology arises out of scalar difference, but every act of scalar mediation, every resolution of difference between surfaces, is a further differentiation, an embedded ecological act.

Scalar mediation affords access to textures that cannot be directly resolved in relation to a single surface, and therein lies its power. When two surfaces are resolved, details not available to ordinary perception are appended to them. The virtual space between them, the medial interstices, are thus harnessed as further intensities, further impulses toward movement. For Boeke, this involves the extraction of a potential for scalable pacifism, a utopian social movement of aggregation and diffusion, a cascading resolution of violence and strife at successive scales. The reader of *Cosmic View* is constituted as a subject, as a perceiving assemblage, capable of interfacing with larger and smaller textures and forming alliances across scales, within the interstices of the book's stabilized scales. The impulse for this trans-scalar potential is created, hopes Boeke, through the virtual movement of resolution itself, a movement produced in the thought of the reader through the *difference* in resolvable texture between each pair of surfaces, a movement that takes place on a larger plane of immanence (to invoke Deleuze and Guattari's concept of the abstract space of becoming).

As we have seen, the filmic versions of the cosmic zoom from the 1960s and 1970s progressively occlude the textural differences between their surfaces, which nevertheless remain real and materially detectable. As one image plate (scalar surface) is dissolved into another and new textural details emerge, a series of artworks is encoded as a spatial contiguity—that is, the images are composed with each other on a singular conceptual plane, through a processual spatialization of time. The primary scalar relation thus shifts from that between medial surface and primary surface (as in Boeke's *Cosmic View*) to that between medial surface features as compositional contiguity. This second arrangement occludes the primary surfaces. The scalar relationship between 10^{-10} (the atom) and 10^{13} (the solar system) in the Eameses' 1977 film, for instance, is not an intensive difference, arising from the conjoining of two separate mediations in the form of (drawing 1 : atom) $\rightarrow$ (drawing 2 : solar system), where "$\rightarrow$" indicates an active perceptual shift from one stable state (scalar relationship) to another. In the 1977 film, this relationship is rather one of a collapsed differential between two points on a surface of resolutions derived only from its medial surfaces (artwork 1 $\rightarrow$ artwork 2). The relation here is between two points within the film, two images in time connected in a contiguous medial space. This is, of course, a real differential, but only between two points or moments *within* the medial unfolding of a secondary surface (the screen). As we have seen, any scalar difference between these two milieus has been occluded.

Recovering scalar difference by revealing the shifting surface negotiations throughout *Powers of Ten* would require interrupting the interlocked space-time unfolding that constitutes the film's smooth cosmic zoom. This is more than can be expected of any casual viewer, for whom the ultimate effect of the film is likely to be a mesmeric reduction of differential resolution across planes and its replacement by a flat ontology consisting of a single medial surface whose textures are distributed as telescoping, contiguous space rather than as a series of scalar negotiations. As the interstices disappear, so do the primary surfaces of the environment, giving rise to what Fredric Jameson theorizes as "a mutation in built space itself" that envelopes the observer and removes all topological and geographical cues that would afford the purposeful navigation of that space or a reconstruction of its hierarchical logics.[31] Jameson calls this "hyperspace." Similarly, Jean Baudrillard approaches this postmodern configuration of space as the "hyperreal," or "a programmatic, metastable, perfectly descriptive machine that offers all the signs of the real and short-circuits all its vicissitudes."[32] While both hyperspace and the hyperreal emerge at the nexus of a complex set of historical trajectories that converged in the late twentieth century, an essential determinant in both cases is scalar collapse at the level of mediation, understood ecologically as a perceptual reduction of available affordances within an environment.

Media, then, participate in both primary scalar difference, or scalar becoming, as they produce virtual space and enable new trans-scalar encounters, as well as secondary scalar difference, or the scalar comparison of stabilized entities across stabilized scalar milieus. Secondary scalar difference is being across difference. Primary scalar difference is becoming. Together, both facets of mediation produce a multiscalar ecology that we, as subjects, inhabit.

LOST IN TRANS-SCALAR ECOLOGY: *POWERS OF TEN INTERACTIVE*

I now wish to turn our attention to a curious, manic instantiation of the cosmic zoom that—deliberately or inadvertently—enacts a mode of mediation that operationalizes both primary and secondary scalar difference, not only at the level of content but at the level of interface. If the film *Powers of Ten* represents the apotheosis of the spatial-contiguity model of scalar mediation and its resulting linear trajectory through a smooth scalar continuum, its remediation into an interactive CD-ROM in 1999 best illustrates its potentials of implosion. *Powers of Ten Interactive*, assembled from materials collected and created over several years by Charles Eames's grandson, Eames Demetrios, continues the curiosity-cabinet curating of the Morrisons' book and pushes it to nearly absurd lengths.[33] Demetrios's concept for the disc was to multiply six thematic "strands" with forty-four scales, yielding 264 intersections, or "stations." Each station contains multiple photographs, video clips, and textual documents. The "Space" strand is essentially the 1977 film, now discretized by scale, shown one tiny chunk at a time. The "Time" strand produces a second discretized narrative of a woman spraying water into a carafe. Each spatial scale corresponds to a temporal scale, from the movement of individual atoms to the entire history of the universe. The "Tools" strand examines tools that humans use to access objects at the given scale, from particle accelerators to language and literature. The "People" strand offers a collection of knowledge workers (mostly scientists, but also a gardener, a fisher, and several writers and artists) whose work engages the given scale. The "Eames" strand is essentially an archive of works or documentations of works by Ray and Charles. The "Patterns" strand documents various symbols that occur across contexts and cultures that visually resemble patterns found at different scales. The effect of aggregating this disparate material is an explosion of content that marries two trends of 1990s digital media: multimedia hypertext and the CD-ROM encyclopedia. Throughout the mid- to late 1990s, multimedia hypertext became increasingly common as a software commodity, as exemplified by Eastgate Software's line of hypertext fiction (delivered at first via floppy diskette and later on CD-ROM) and multimedia CD-ROM magazines such as *Launch* and *Blender*, which used the relatively high storage capacity of the optical-disc format to deliver hyperlinked text, images, audio, and video clips. The same era and technology saw the rise of the CD-ROM encyclopedia, as exemplified by *Compton's Multimedia Encyclopedia*, Microsoft's *Encarta*, *Grolier's Multimedia Encyclopedia*, and the *World Book Encyclopedia* CD-ROM.

Powers of Ten Interactive, developed starting in 1994 and finally released in 1999, aspires to be both a digital encyclopedia of scale and a hyperlinked archive of the Eames legacy. Like multimedia encyclopedias, this disc contains a vast archive of data, but unlike them, the archive is not organized according to any single high-level indexical layer. There are no individuated articles or essays beyond the textual fragments in the "digital anthology," and even the six primary thematic "strands" foil any attempt to navigate them linearly or to access any top-level index of their contents: to navigate a strand is to be continually halted, presented at every scale with a bewildering array of media and branching paths. Similarly, if one attempts to navigate the disc via a linear progression of scales, one is quickly lost down a branching rabbit burrow of links, interviews, images, and textual documents that thematically and hypertextually lead to many other, noncon-

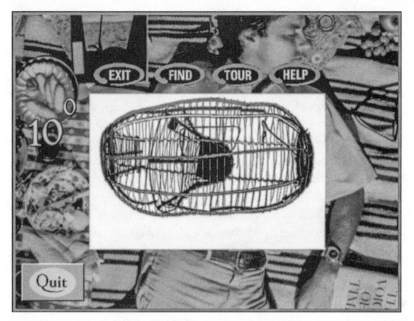

Figure 22 ▸ Giant Fishtrap interface from *Powers of Ten Interactive* (1999).

tiguous scales. There is little to guide the user, and the result is inevitably a kind of personalized journey through a vast mash-up of scalar media with no readily discernible organizational logic. On this disc one does not "look up" information on a given subject, but rather generates a path through the chaosmos of scale according to momentary choices based upon singular affinities—or blind chance. The disc's only navigational overview is a page referred to as the "Giant Fishtrap" (figure 22). Six more-or-less horizontal, color-coded lines (representing the strands) are intersected by more numerous vertical lines (each representing a discrete scale), suggesting the woven structure of a traditional fish trap of the sort utilized by the indigenous peoples of North America. Each point on the mesh represents an intersection of scale and strand, and clicking on it takes the user to the "station" that occupies that intersection. The Giant Fishtrap, one of the first concepts generated by Demetrios during project development, thus serves as the disc's inspired emblem of its own rhizomatic structure.[34]

The Giant Fishtrap is a differential scalar diagram. As hypertextual scalar cartography, it forgoes spatial contiguity in favor of an interwoven matrix of intersectional scalar potentials. In this sense it is intensive: the Fishtrap itself indexes the totality of the disc's affordances as its own surface of difference, without positing any set of stabilized objects with which to directly interact. As interface it highlights the interactive potentials of the disc as distances within a matrix of different surfaces. As soon as one clicks somewhere on the Fishtrap, both the singularity of that intersection and its deeply hyperlinked topological position become immediately apparent. In the chaotic flux of *Powers of Ten Interactive*, disciplines and knowledge structures are linked to technologies of scalar access to artworks to patterns to timescales in a potentially endless loop through the scales of the universe. Each scale links to many others, but only through differentials mediated by thematic strands. Here, finally, is a form of scalar mediation that stabilizes its scales as hypertextual nodes. Each scale captures singular processes, but those processes interact across all scales, producing the differentiation that we experience as culture, tools of intervention, objects within milieus, and so on. Ultimately, *Powers of Ten Interactive* generates a decentralized structure in which each node (stabilized scale) connects differentially to a vast number of others through various associative links and the elaboration of various trans-scalar dynamics. (I will discuss more of the interfacial elements that generate these links shortly.) In a network graph visualizing the explicit interscalar connections of the work, its rhizomatic structure becomes immediately apparent (figure 23).[35]

Actually navigating *Powers of Ten Interactive* is a kind of meta-exploration: one explores the Eameses' exploration of modern culture through design, explores the musings of various scientists exploring the empirical nature of objects at disparate scales, explores the images produced by artists exploring various intensities discovered as virtual potentials in their world, explores the very medial apparatuses that have afforded access to trans-scalar surfaces and thus enabled this quixotic, kaleidoscopic, yet manifestly and necessarily incomplete compendium of scalar knowledge.

If the Giant Fishtrap serves as the emblem of the disc's rhizomatic structure, the ordinary station interface (figure 24) acts as a dashboard that affords the actual navigation of the work's virtual surfaces. While a tiny version of the Fishtrap returns on every screen, it is too small to read as a map. Each station contains multiple "images," some of which are video clips, navigated via arrows and represented by a timeline that is curiously not scaled to the length of each slideshow. Clicking on a crosshair icon converts the currently displayed image into a contextual image map; subsequently clicking on hotspots within the image may yield pop-up dialogue boxes with further information about that element. There is, however, no indication of when an image contains such affordances and when it does not. Captions can be turned on or off by clicking on an image while not in crosshair mode. Other icons bring up a text overview that explains the theme of the current station and a digital anthology of related texts

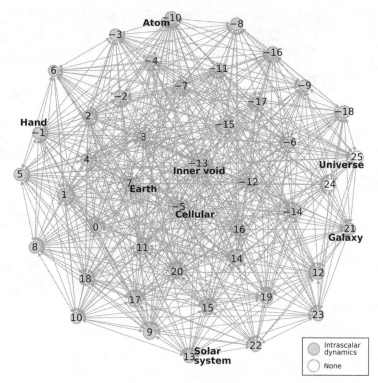

Figure 23 ▸ Network graph of interscalar connections in *Powers of Ten Interactive* (1999).

stored on the disc (interview transcripts, book chapters, etc.). To the left of the central image is an indication of which scale one currently occupies. Hovering near this indicator brings up further navigation controls through which the user can change stations, moving up or down the scalar spectrum one scale at a time—but the action one selects by clicking is deferred until the user moves the cursor off of the scale indicator. Color-coded boxes on the right side of the screen link to the six strands as they intersect with the current scale. At the bottom of the screen are three links to different stations (strand-scale combinations) that are processually, disciplinarily, thematically, or aesthetically related to one or more elements that appear somewhere within the current station's images, clips, and textual repository. Clicking on one of them transports the user to a completely different scale and (usually) strand. These three links are different for every station, thus multiplying the outgoing links from any given scale, but only rendering them visible in relation to their respective strands. Finally, three oval hyperlinks in the corners of the screen change contextually, based not upon the current intersection of scale and strand, but rather the path the

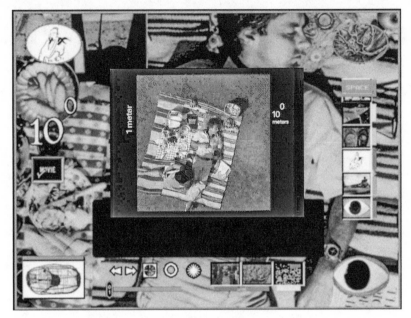

Figure 24 ▸ The "station" interface from *Powers of Ten Interactive* (1999).

user has taken to arrive there. Clicking on one of them jumps the user back one, two, or five stations in their browsing history.[36]

The effect of this almost absurdly cluttered interface, with its incomprehensible icons and bewildering number of options, is to layer a superstructure of imprecision and misdirection on top of the disc's rhizomatic but disciplined structure. Yet, far from foiling access or obscuring the potentials of this structure, I wish to argue, the sheer madness of the interface extracts its maximum potential. Three elements contribute to the effect of imprecise misdirection. First, the design, which crams every function onto a single plane without hierarchy or metadirection, effectively severs the link between form and function. It is surely the most un-Eames-like design that could be devised![37] Second, the interface's temporal delays—resulting from lags in loading content from a slow optical disc, limitations that prevent clicks from being registered while video is playing, and design quirks such as the aforementioned delay in switching scales—introduce constant deferrals that can and do escalate into both loops and random insertions into unintended stations.

The third feature that contributes to the interface's uniquely imprecise misdirection is its resolution: the entire screen is software-limited to a resolution of 640 by 480 pixels and a color depth of only 256 colors. *Powers of Ten Interactive* is thus the lowest-resolution instantiation of the cosmic zoom that we have yet encountered, which of course means that it amplifies the resolution-based effects of scalar mediation: since changing scales means resolving new surface detail, an incredibly low surface-resolving power requires a correspondingly large scalar differential. At the same time, the dense layout packs a minimum of 283 hypertextual links onto every 0.3-megapixel screen (or more, depending on the number of contextual image-map hotspots); given the low resolution, the interface can scarcely resolve enough distinct cursor positions to render them all discretely accessible. When we add in the effects of temporal delays and a flat topology of affordances, any purposeful navigation on the user's part is continually foiled. The informational topography of the software thus becomes a kind of trans-scalar wilderness, a milieu of milieus. Because the disc's contents index and model cultural mediations of scale, and thus scalar access techniques, this interfacial myopia helps to emphasize the scalar differentials (trans-scalar environmental intensities) that one faces when attempting to stabilize a milieu by traversing its surfaces. The program thereby models not only secondary scalar difference as environmental (exploratory) mediation, but also something of the ontogenetic potential of primary scalar difference in the emergent gap between ordered human intentionality and rhizomatic scalar irruptions.

Sadly, *Powers of Ten Interactive* was soon obsolete; by the end of 2001 (a mere two years after release) it no longer ran properly on newer computers, and it quickly went out of print.[38] Demetrios and the Eames Office continued to recycle some of its content online and in a 2008 video compilation, *Scale Is the New Geography*, intended as a set of educational resources for the teaching of the original 1977 film. A more ambitious website was launched in October 2010, with a simplified interface: a vertical scalar axis on the left of the screen and, for any given scale, several strands to the right (only a few scales now linked to all six). The scale-strand intersections, moreover, offered fewer images and accompanying text, without video or sound. A pale shadow of the CD-ROM, the site largely omitted the features I have highlighted. Gone was the sense of navigating scalar difference in all of its rhizomatic complexity, lost in ecological detail rather than mastering a linear course, guided by an overview that collapses scalar difference. The entire site was taken down in July 2014.

MEDIAL ECOSYSTEMS

The navigation of scale has become a central theme and mechanism of our contemporary culture, and scalar media serve to present as cultural surfaces those mediating techniques that enable and hinder the navigation of the many other surfaces of our milieus. Their mediations function primarily by affording new means to access and reorganize the techniques of scalar mediation itself. Intentionally or unintentionally, *Powers of Ten Interactive* acts as a scrambler of codes, a short-circuiting of the scalar boundaries of disciplines, tools, and artistic representation. As mediation itself is scrambled within its interface, difference is preserved as a chaosmos from which scalar intensity can be extracted. The navigation of scale, this CD-ROM not only reminds us but forces us to experience directly, is a messy, combinatoric process. There can be no scalar overview but only local intensive differences within a nevertheless irreducible web of interdependencies. The fact that we can only resolve the detail of a surface one level at a time, combined with the knowledge that differential processes span all scales and are linked in vast interdependencies, has profound ethical implications. Any entity—embedded within and part of the fluxes of a trans-scalar environment—that collapses scalar difference through medial practice thereby turns mediation against its primary role of differentiation. The result can only be a reduction of affordances for entities across the scalar spectrum. That is, any increase in the extensive expanse of scalar access, without an attendant increase in ability to resolve difference, can only do violence to the larger trans-scalar ecosystem that subtends all encounters between surfaces.

It remains a vital necessity, then, to critically engage our media, to ask how they do their work of stabilizing certain scales and enabling or constraining our access to trans-scalar processes. Figure 25 gathers the network graphs of eight cosmic-zoom projects (two of which I will discuss in the following chapter) into a kind of menagerie of medial systems. Each of the graphs was generated automatically using the same layout algorithm, allowing for direct comparison of their constraints and affordances as scalar media. These are the shapes of scalar affordances. If, as a group, they suggest a micro-ecosystem of animated creatures, this is entirely appropriate. Every mediation enables certain forms of trans-scalar access and forecloses others by tracing a model of scalar difference. Every medial negotiation between surfaces implies a form of trans-scalar ecology.

How, we might ask of each medial organism, does it model interactions *within* scales, and thus demarcate scalar boundaries and the knowledge domains keyed to resolve such detail? How does it model interactions *between* scales by constructing connections, epistemic networks that enable or constrain the types of trans-scalar processes we are able to conceive of and partake in? How ecologically robust is its model of scale? Is it resilient

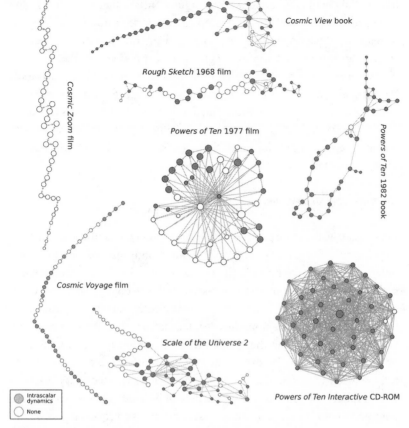

Figure 25 ▸ Comparative menagerie of cosmic-zoom network graphs.

(redundantly connected) or vulnerable to the severing of edges through scalar myopia? Is it hierarchical (with certain nodes disproportionately central) or rhizomatic (containing many edges, well distributed to produce a high degree of average connectedness)? Are its scalar nodes alive with the Brownian motion of intensive differentiation (intra-action), or do they serve as mere inanimate scaffolding for a larger scalar hierarchy? Medial evolution, like organic evolution, doesn't necessarily progress toward optimization, nor, certainly, toward more egalitarian or creative community building. Each trans-scalar being possesses a different navigation system, and each is capable of seeing certain details in the world while remaining blind to others. Each is a functioning system of affordances. Which, I wonder, would you prefer to meet while swimming in the murky waters of the Anthropocene?

Such relational graphs are of course limited in their own resolution of differential detail and cannot suffice to fully answer these questions. The power of graphing scalar relationships in network form, however, is that instead of merely describing the empirical interactions of an observed system, we are here attempting to graph the expressive *potential* of scalar models to resolve difference, and thus their efficacy and robustness as tools for navigating and creatively interacting with the surfaces of the world in all of their manifold detail. Scalar media that enable their users to access the singularity and vitality of other scales prime them for these encounters, opening new potentials for resolution, new senses, and new ways to conceive, encounter, and become alterity itself.

WHY DOES SCALAR DIFFERENCE MATTER?

I have been elaborating two related forms or procedures of scale. The first, or primary, form of scalar dynamics is one of intensity—scale as internal differentiation, as difference itself in its act of differing. The second form of scalar dynamics is one of extensity—scale as the resolving of alterior detail, as mediation. Both dynamics fundamentally determine our experience of being in the world. It is nonetheless vital that we understand the second form to be derived from the first. Difference is primary and underlies the process of mediation; mediation does not come to a table that is already fully dressed with stabilized objects. Mediation is a secondary elaboration of differentiation that composes new objects and new subjects by attempting to resolve one surface against another. As a process it partakes in the *production* of detail through its differential resolution. When we view scale from the inverse perspective, as a mere set of size relationships between entities assumed to be stable and to have existed prior to their scalar elaboration, we relegate scalar difference to a secondary role. This is, naturally, the tendency of the empirical sciences, which typically frame their own disciplinary work as a process of discovery rather than one of co-creation or composition. What this cannot account for, however, is becoming, the swerve. Why does each act of becoming seem to produce a new scale, with new rules of behavior?

The solutions that have most readily presented themselves are those of Newton or Einstein, who assumed that chaos—or different laws for different scales—is only a surface-level effect of a more profound, more elegantly simple underlying structure. This approach to scalar difference is embodied aesthetically by the Eameses' films and socially by the universal equivalence and resulting cultural structuration of capital but, as Morrison indicates, becomes empirically untenable in the era of quantum mechanics.

We should remember, however, that if quantum mechanics recognizes the fundamental role of chance in the elaboration of difference, and thus the singularity of all scales, this is not a radically new conception of difference but a return to a pre-Newtonian world that would be quite recognizable to the likes of Lao Tzu, Heraclitus, or Lucretius. This indicates that quantum mechanics is not necessary as the platform of thought from which to consider scalar differentiation as a primary form of becoming that engenders difference rather than conforming to similarity. We may turn also to disciplines such as the arts and humanities, which are perhaps more comfortable with the ongoing elaboration of difference rather than the pursuit of universality.

Moreover, just as the swerve of scalar alterity is not a concept new to modernity, Newtonian epistemology and aesthetics are merely elaborations of the milieu-building tools employed by the authors of such treatises on scale as the Torah, the Gospels, book VII of the *Republic*, the Quran, the *Summa Theologica*, and *A Brief History of Time*. These metaphysical and aesthetic battles are not likely to be laid to rest any time soon. Works and systems of scalar mediation represent the front lines in this war, which frequently sees new innovations of armament on both sides.

The distinction between primary and secondary scalar difference can help formalize an alternative solution to the scalar paradox raised in the introduction—namely, that scale appears to be both an ontological determinant and an arbitrary, socially constructed set of divisions. Rather than collapsing one side or the other of the paradox, we need to understand that there are two types or orders of scalar difference. In the account I've developed here, primary scalar difference, or "scale from the inside," is becoming itself, the elaboration of matter into new assemblages, which in turn produce newly differentiated systems. Scale is a primary form of differentiation, baked into the universe, as it were. In this sense it is certainly anterior to the human. All forms of consciousness arise only through scalar difference, as one mode by which matter-energy folds back onto itself in a kind of encounter: knowing itself. But in this epistemic phase, as practiced by humans, scale must be mediated by conceptual and technical apparatuses. These have the effect of reifying objects and resolving particular milieus. Humans—and other entities—make resolving cuts into the manifold universe in order to isolate certain features. These cuts are indeed arbitrary and socially constructed, but no less necessary for that. This is the realm of secondary scalar difference. We initiate it as mediation, and it is entirely contingent—we could do it differently.

As cosmic-zoom media demonstrate, there are profound political, epistemological, and ontological implications to how and where we make our resolving cuts. Nonetheless, wherever and however we make them, they serve to reveal primary scalar difference. Our media contain and code that difference, but wherever we look, we will find it. We make scales, but scale unmakes us. The paradox, then, is really a spiral: we isolate scales and then find difference teeming on and across those surfaces. Disciplined knowledge production contains and orders that difference into newly fixed objects and scales. This reveals more irruptions of nonhuman scalar difference. And so on.

This process of scalar mediation produces new objects and new subjects. At the same time that humans have come to understand themselves as a geological force—that is, as willing or unwilling trans-scalar milieu builders—the Anthropocene has produced radical innovations in the domain of scalar mediation. The human subject is constituted by scalar mediation in processes that foreclose certain potentials of trans-scalar action (for example, dissolution of the individual as primary unit of organization and identity, abdication of control over nonhuman entities, establishment of communal forms of ownership without an autocratic state) and enable others (global climate change, systematic destruction of biodiversity, globalized markets). How we approach scalar alterity, how we mediate scale, constrains how we constitute scalar problems and, thus, which potentials remain open for their solution.

Our critical project remains, then, to recover scalar difference without rejecting robust scalar mediation. The solution is not to retreat to a single scale but to approach scalar differentials differently. This entails affirming an intensive process of scalar mediation, not an immediacy of trans-scalar access or the bracketing out of scale in the mistaken view that this will eliminate scalar hierarchies. We need more scale, not less, even as scale mania grips us. This is provided, of course, that scale becomes an engagement with and opening to alterity and not its occlusion, collapse, or colonization. In the final chapter of this book I will explore the current state of cosmic-zoom media alongside the question of the digital. How does our digital milieu mediate scale differently? What has it drawn from the legacy of the cosmic zoom, and what is its unique contribution? How might the lens of scale shed light on the co-construction of media and subjectivity in the age of the pixel and the database? As we examine the foreclosures as well as the generative potentials opened by these new forms of scalar mediation, we would do well to keep in mind tendencies we have uncovered in the genealogy of cosmic-zoom thinking, retaining and recovering those dynamics of scalar difference that bubble up as resolution between surfaces in Boeke's *Cosmic View*, self-referential temporal difference in the Eameses' *Rough Sketch*, resistant quantum uncertainty in the *Powers of Ten* book, and the interfacial chaos opening onto rhizomatic (anti-)structure that characterizes *Powers of Ten Interactive*. We shall need all of these tools and more as we navigate and modify the milieu we have created as a species.

a digital universe?

Database, Scale, and Recursive Identity

HOW BIG IS "BIG DATA"?

The cosmic zoom has served as our guide to a transdisiplinary theory of scale that accounts for the paradoxical nature of scale as both an ontologically primary form of difference and an arbitrary social construct. This final chapter seeks to apply scalar difference theory to contemporary digital media. In particular, I explore the database form as a dynamic of scalar mediation with profound effects on human subjectivity. Here, then, we will return to the human subject and ask not only how scalar dynamics contribute to its construction, but how scale may point the way toward alternative subjective formations that may break free of the scalar constraints so vividly embodied by the film *Powers of Ten*, explored in chapter 4. My intent here is a speculative reconsideration of human subjectivity from a new angle. This is not to reinscribe the human subject in a central role; from the outset, this book has argued for a nonhuman understanding of scale. The question, rather, is whether subjectivity can be approached differently, from outside, as it were—from the perspective of scalar difference. Contemporary database-driven media will provide us with just this opportunity. Throughout this chapter I examine a number of more recent cosmic zooms and analyze ways that the database form has fundamentally altered both their visual modalities and scale-stabilizing functions in the larger digital media ecology.

Powers of Ten Interactive, the bizarre hypertextual instantiation of the cosmic zoom that I discussed in chapter 5, has already edged us into the affordances of the digital. Yet my argument in this chapter is not that digital media naturally, automatically, or even commonly enable a more densely networked topology of scalar articulation. We saw, in the case of Kees Boeke's *Cosmic View*, that analog media can engage an extremely thick ecology of inter- as well as intrascalar detail. As I argued in chapters 3 and 4, if *Rough Sketch* and *Powers of Ten* elide such interrelationships, this is not a necessary analog limitation so much as an example of analog media occluding their own seams, or the scale-discreteness that constitutes them. But just as analog filmic media must occlude their own inherent seams in order to produce an analog model of scale, digital media—as I hope to show here—do not necessarily model a discretely articulated environment (digital scale). When the affordances of the discrete pixel and the discretizing database are jointly harnessed in digital cosmic zooms, the apparent result is more often than not an *analog* model of scalar ecology. Yet as we shall see, the possibility of a truly digital model of scale remains open when digital scalar mediation is mobilized recursively.

Pixel and database. These are the two poles of digital mediation that ground this chapter's foray into digital scalar mediation. We might think of these as the interfacial layer and the structured data layer of digital mediation, respectively, but I mean to complicate these distinctions in the process of highlighting their relation to scale. I do not mean to reduce digital media to only these two elements, of course, nor to elevate them to the status of transcendental ur-categories. The pixel is only one form of digital representation, and the database is only one form of structured data, just as scale is only one of the ontological territories mediated by technical and conceptual systems. Each, however, bears a special relationship to the mediation of scalar difference, which constitutes the study at hand. The pixel is the basic building block of the digital surface, a discrete unit of spatial representability that serves as a hard limit on the resolving power of a digital media system and an infinitely iterable, coded assemblage that functions in its very liquidity as the primary means by which scalar transformations occur within contemporary media systems. The database is, conversely, a homogenizing system of structured lists and protocols that serves to sort, categorize, store, and retrieve detail as data. As we shall see, it is the liquid pixel that most effectively occludes scalar difference, and the seemingly rigid database that opens digital mediation once again to the potentials of the trans-scalar encounter with alterity itself.

We might call this the scalar paradox of big data. If big data is "big" in the sense that there's a lot of it, what exactly does this mean? That the information available to computational systems and their human managers affords greater access than ever to the ecological detail of a given milieu of data collection? As we have seen, detail can only be accessed through processes of scalar stabilization, which require resolving cuts. Having massive "quantities" of data does not, then, automatically grant unfettered access to granular detail. No matter how much data is collected, its access is still subject to the limits of resolution I elaborated in chapter 2. Quantity, then, cannot explain the "bigness" of big data.

Alternatively, is big data "big" because is encompasses so many domains of digitization and control? Danah Boyd and Kate Crawford contend that "Big Data is less about data that is big than it is about a capacity to search, aggregate, and cross-reference large data sets."[1] As they note, big data is really an aggregate of technological infrastructures (of data collection, processing, and algorithmic manipulation), analytical methods (making claims based on analysis of large datasets), and mythic beliefs (that large data sets are somehow objective, accurate, or truthful). The cultural currency of big data is clearly due to the efficacy of this triumvirate, which has led to its spread to many, if not most, domains of cultural activity. Yet while data collection, algorithmic control, and positivist narrative do indeed encompass many domains of activity and disciplinary knowledge formation, this does not necessarily entail the coupling of these systems. In fact, the operationalization of mass data is subject to the same domain boundarization that governs nondigital methods of information gathering and processing. Put another way, data doesn't mean anything until it is interpreted, and interpretation remains the purview of the sort of disciplinary formations I explored in chapters 4 and 5. In this sense, disciplinary scalar optics precede data operationalization. Big data isn't big until it is interpreted (operationalized) as "big," and this is a function not of domain-crossing but of domain definition. The parameterization of data collection (what we might call its thickness or depth) may expand, but parameters must pass through disciplinary optics and become operationalized before they are articulated in any way. Disciplined knowledge transforms the world into potential data points, so we might just as well say that we have "big discipline."

Finally, we might say that data bulks up as it gains trans-scalar potential. We could track this feature as the number of scalar thresholds it spans or the density of feedback loops it operationalizes. Or more properly, in the terms I have been proposing in this book, big data is big in relation to the possible scalar articulations it enables. This is an interfacial as well as procedural distinction: big data as a form of scalar navigation embedded into the milieu of its denizens. Here the question is not how much data is collected or ready at hand, or how many domains of experience it encompasses, but the resolving cuts that it either enacts or makes possible.

Big data is a method of scalar access, and thus a medial dynamic. As a conjoined way of seeing and knowing, mass data collection animates NASA's deep space telescope images and mediates our understanding of the cosmos,[2] drives the CIA's real-time tracking of suspected terrorists, creating a vertical mode of representation,[3] mediates our understanding of endangered species,[4] and generates a datafied version of human social media users that is visible to us only as an array of targeted advertisements and customized shopping services through which we are "manipulated and controlled, made homogeneous as entries in lines of a database or records list."[5] In a big data society, everything is subject to datafication, surveillance, and analysis. What emerges, however, are a plethora of new entities at new scales.

This chapter proposes a new analytic of digital media based upon scalar alterity. Implicitly, then, it offers a new critique of big data that casts it neither as a boon for optimization, regularization, and command and control, nor as an inherently inauthentic and colonizing caricature of human identities—though it may well also serve both of these functions. Rather, I read big data as a form of scalar differentiation that opens up certain forms of trans-scalar encounter and forecloses others. If my exploration of the cosmic zoom in this book led up to the previous chapter's articulation of a transdisciplinary theory of scalar difference, this chapter attempts to apply that theory to a contemporary context of fraught scalar negotiation within an ubiquitously digital techno-social milieu. This is the self-help section of the book. But in the context of scalar alterity, what is the self?

UNIVERSAL PIXELS

In 1996 the National Science Foundation released what may still be the most ambitious and technically advanced version of the cosmic zoom yet, *Cosmic Voyage*. This half-hour film, directed by Bayley Silleck, was shot and (for its digitally simulated sequences) rendered on IMAX film. The producers hired Hollywood special-effects teams at Pixar and Santa Barbara Studios to work with simulations developed by the National Center for

Figure 26 ▸ Quarks as meteors. Frame from *Cosmic Voyage* (1996).

Supercomputing Applications (NCSA) at the University of Illinois. The film transitions between live-action scenes and computer-generated ones, while its narrator, Morgan Freeman, muses on the importance of curiosity and discovery for the history and future of the human species. The first of several cosmic zooms begins in Saint Mark's Square in Venice, Italy, where Galileo purportedly first demonstrated his telescope, enabling views of distant objects (according to Freeman's narration). The sequence jumps scales for the first several powers of ten (filmed in live action), then begins a long, computer-generated zoom-out to the edges of the known universe. "What lies beyond this cosmic horizon, we cannot see, and do not know," proclaims Freeman solemnly, and the setting changes to Delft, the Netherlands, where Anton van Leeuwenhoek produced the first effective microscope and "discovered a living kingdom inside a drop of water." Here a child peers into a water drop quivering on a leaf, initiating a computer-generated inward journey through strands of DNA and eventually into a field of quarks, which zip around like angry little meteors. The animators have taken the dubious step of rendering these massless entities as particles leaving trails of light blazing behind them (figure 26).

While *Cosmic Voyage*'s narration emphasizes the production of scientific knowledge and the scale-resolving technologies that enable it, the film never mentions specific disciplines by name. Even more significant from a medial point of view is the reduced variety of source artwork in relation to earlier cosmic-zoom films: *Cosmic Voyage* forgoes still photographs or schematic representations, relying wholly on cinematic optics and computer-generated imagery (CGI). The IMAX-filmed sequences

Figure 27 ▸ Two galaxies collide. Frame from *Cosmic Voyage* (1996).

make use of macro lenses or elevated platforms (aerial cinematography), cinematic technologies rather than scientific instrumentation—despite the narrative structuring devices of the telescope and microscope. The computer simulations similarly bypass the intermediate image production (diagrammatic or photorealistic) of the Eameses films and book, in favor of a perfectly continuous visual field, a smooth disciplinary surface, and of course a smooth scalar contiguity.

Even in *Cosmic Voyage*'s most spectacular visualizations—such as an operatic sequence in which two galaxies slowly collide (figure 27), exchanging luminous particles in a cosmic vortex both delicate and violent, and producing an entirely new assemblage of swirling clusters—each simulation is clearly made using the same particle renderer used for sequences at other scales. Though we are informed that the galactic collision unfolds over "a billion years" in real time, its individual stars look like electrons inside the simulated atom, or the fiery explosion of a comet hitting the earth in another of the film's sequences. Thus, while the use of a particle renderer to visualize NCSA's database of dynamic objects allows the filmmakers to eliminate the visual scalar seams I discussed in chapter 3, this comes at the cost of homogenizing scalar difference into generic dots on the medial surface. Different scales (still marked by powers of ten and delineated by superimposed quadrats, this time circular instead of square) not only bleed into each other on a smooth, contiguous surface, but look exactly alike. This is a new take on scalar collapse: different events depicted at radically different scales are clearly composed of the same algorithmically animated particles. As Philip Morrison might say, "We are now entering the realm of universal pixels."

Graham Harwood notes, in his brief rumination on the pixel, that it "has become both a mirror and a lens, reflecting and shaping the realities of its own making."[6] The pixel is the smallest unit of resolution in a digital imaging system, but also sits at the threshold of the digital and the analog, and in fact constitutes their point of translation. Any threshold between the analog and the digital is nearly always also a site of trans-scalar encounter, and thus scalar mediation. Moreover, the physical, protocological, and symbolic interfaces necessary to effect radical domain transfers such as that between the human sensorium and structured data not only enable but actually necessitate the medial stabilization of scalar difference in the manner I have elaborated throughout this book. The translation between the digital and the analog, however, can conceal as well as stabilize and reveal. *Cosmic Voyage*'s use of simulated imagery suggests that we are in some sense experiencing the underlying data collected by NCSA, the University of California at Santa Cruz, and the Smithsonian Astrophysical Observatory, though the source of such data is never named in the film itself, in the narration or in titles.[7] Neither is the process by which such data was obtained nor the processes by which it has been visualized ever reflected upon. Of all the cosmic-zoom projects I have discussed in this book, *Cosmic Voyage* is the least self-reflexive, despite the fact that it was ostensibly produced for the purpose of promoting science education.

Cinematic in technique and nondisciplinary in content, what is visualized here is not the immediate universe or disciplined knowledge production, but a homogenized, self-congratulatory version of Science. "From the beginning, we were explorers, inventors, and technicians," Freeman tells us, but none of these activities is on display in this visually mesmerizing dance of pixels. Tellingly, the computer-simulated sequences were processed through a specially developed visualization technique dubbed Virtual Director, which allowed the virtual camera operator to navigate through the scientific simulations from a first-person perspective.[8] This determined the rendering path of each sequence, ensuring that it conformed to the smooth aesthetic of Hollywood's great technical innovation of the previous two decades, the Steadicam. The Smithsonian's confused press release highlights the elided relationship between representation and knowledge production with unintentional irony when it declares that "key members of the production team developed a new technology called the 'Virtual Director' to help simulate the computer-generated images, a process which is enticing Hollywood filmmakers."[9] In this telling, the images themselves are simulated, rather than the underlying datasets or the cosmic phenomena that ostensibly constitute the film's subject matter. The digital becomes a medium not of simulation (which requires the modeling of scale- and discipline-specific interactions) but of scalar homogeneity.[10] Thus while Virtual Director is an interface, a means to narrativize and singularize the navigation of a database, it also demands its own ontological priority. The scientific datasets and theoretical models that drive the primary simulations are conceptually erased in favor of a second-order simulation: that of a scale-invariant point of view, a universal perspectival equivalent that aggregates and dissolves imagery from all scales into a flattened visual topography.

In this translation process, the pixel acts as a smooth surface, a frictionless conveyance between stabilized scalar domains—which of course it helps to stabilize in the first place. In cinematic terms, the virtual space of the pixel's potential transformations is expressed ontologically as the "morph." Vivian Sobchack argues that "there is nothing quite so unheimlich as the sense the morph conveys that, in its totalizing achievement and containment of change, the flux of transformation is itself undone, become essentially meaningless—not only because this change is endlessly repeatable, but also because it is essentially reversible." Thus *Cosmic Voyage*'s particle simulations, shifting scales through the unfolding of cinematic events, do not ultimately signify transformation at all. Because the digital morph "empties change of its temporal specificity and symbolic and existential value," *Cosmic Voyage*'s boundary-spanning pixels not only affectively unravel the representational changes they effect, but also essentially nullify their scalar transitions, laying bare the boundary work they accomplish. For Sobchack, the morph asserts "not only sameness across difference, but also the very sameness of difference."[11] With regard to the notion of primary scalar difference I developed in the previous chapter, we can now see that pixellated medial interfaces collapse scalar difference not only strategically, as a form of trans-scalar access shoring up the human subject as disciplined knowledge producer (chapter 4), but even more fundamentally, as an infrastructural enactment of universal transcoding.[12] In this regard the pixel's version of palindromic scalar difference acts as the necessary correlative of the database structure that underlies it.

I further explore the database in the following sections. First, however, I wish to suggest the possibility of writing a new role for the pixel in trans-scalar digital media. Our cinematic avatar for this transformation may well be the "digital multitude" rather than the morph. According to

Figure 28 ▸ The zombie multitude (out)scales the walls of Jerusalem. Frame from *World War Z* (2013).

Kristen Whissel, the "the multitude materializes an extreme version of collectivity antithetical to individuation."[13] Enabled by the same scalar mutability that animates the morph, the digital multitude is a massification that is potentially nonreversible in that it inaugurates a change in subjectivity along with its change in scale. While cinematic narrative tends to contain the digital multitude, ultimately vanquishing it and reasserting the scale-mastering human subject, it represents a threat whose virtual vectors persist as existential dread. Consider, for instance, the zombie film. While the genre-defining films of director George Romero linked the zombie threat to human dynamics of scalar aggregation—racism and consumer capitalism in *Night of the Living Dead* (1968) and *Dawn of the Dead* (1979), respectively— the advent of the pixel has rendered the zombie beyond narrative repatriation. In its digitally massified form, the zombie outscales the human and reveals its future. This clock cannot be run backward, only held at bay for an indefinite period of time. When the posthuman zombie multitude overruns the walls of Jerusalem in a visually stunning CGI sequence in Marc Forster's *World War Z* (2013), the pixel's transformations align with those of the human subject, forever becoming zombie. Here, the human subject, emptied of interiority and recomposed into an undead digital multitude, is not dispersed by circuits of affect or capital so much as it is transformed by jumping spatial scales to the size of a city, and temporal scales to the duration of biblical apocalypse (figure 28). Because the real shifts in human subjectivity (discussed later in this chapter) are being produced by digital technologies themselves, the digital multitude, as catalyzed by the paradoxically mobile pixel, is self-reflexive.

Straddling both the scalar and digital-analog boundary as infrastructural mediator, the pixel exploits its dual nature as both membrane and code. As membrane, the pixel is a gateway through which an image must pass in its journey from analog contour (primary surface) through digital encoding to analog contour (as wave perceived by an analog observer). On this model, the digital is an intermediate step, a wave parameterized as a set of values capable of being stored, transmitted, transformed, and displayed—and thereby ultimately reconstituted as an approximate wave again. The medial membrane is one of digitization, a coded approximation of difference in binary terms. It is also, however, possible to view this interface differently—not as a membrane of encoding and decoding through which the analog passes, but as a primary flux of differentiation that articulates and engenders both digital and analog forms. Instead of capturing analog contours, approximating them in binary terms, and then retransmitting them in altered forms, this second model of the digital posits it as the primary set of forces that produce potential pathways of movement and articulation. Thus, the zombie pixel does not receive the preexisting analog as an input, nor reproduce its facsimile as an output, but rather potentiates the analog as a milieu-function. Rather than representing surface details of an environment, it is a force of scalar differentiation that helps to compose that environment.

In the parlance of this book, we might suggest that this *scalar digitality* is a feature of any milieu that makes resolving cuts possible.

DATABASE, ANALOGICAL STRUCTURE, AND CONTROL

The database wants to forget how to zoom. This is its greatest trick.

Lev Manovich argues, in his classic treatment of the database as media, that database and narrative have been locked in a medial struggle throughout history. Narrative, and its computer science correlate, algorithm, model the world as a series of events. The database models the world as a collection of objects. The medium of the book (codex) supported both forms equally, photography favored the database, cinema swung toward narrative, and computational media privileges the database again. Manovich's characterization of the database itself is essentialist: "As a cultural form, the database represents the world as a list of items, and it refuses to order this list." He contrasts this to narrative interpretation, which constructs an ordered, meaningful sequence, rendering narrative and the database "natural enemies."[14] Most other humanities scholars have followed Manovich's lead, accepting this characterization of the database form as essentially unstructured and nonlinear. Even N. Katherine Hayles, when taking Manovich's database-versus-narrative story to task, takes the tack that the two forms are "natural symbiots" rather than enemies:

> Because database can construct relational juxtapositions but is helpless to interpret or explain them, it needs narrative to make its results meaningful. Narrative, for its part, needs database in the computationally intensive culture of the new millennium to enhance its cultural authority and test the generality of its insights.[15]

In other words, narrative gives data meaning, while the database enables us to scale narrative. These are important insights, but they do not trouble Manovich's characterization of the database form; they only emphasize the continued necessity of narrative in the construction, or reconstruction, of meaning. With Hayles as a notable exception, critical attention in recent years has generally turned toward the algorithm. The database is assumed to be significant but banal, typically treated as merely the structured data form upon which the algorithm conducts its business. This may be how the database appears/disappears to us now, but this is a set of historically and medially contingent features. I propose in this section that a more historicized reading of the database form will better uncover both its past and future potentials with regard to scalar mediation.

Databases were not always, and are not necessarily, either unordered or infinitely appendable, as Manovich asserts. E. F. Codd formalized the relational database model in an extremely influential article in 1970. Existing databases at the time employed either hierarchical or network structures. A hierarchical structure is organized like a family tree: a given entity contains, within its scope, other entities, which branch into other entities, and so on. While this takes the form of a genealogy, it implies the structure of *scala*— a hierarchical ladder structure arranged along a scalar axis. This form of database has the advantage of encoding a set of relationships directly into its structure. It is simple and intuitive, and thus proved to be an efficacious modeling strategy in early database design, taking place within clock-limited and interface-thin computational systems.[16] A hierarchical database encodes real-world relationships into its structure; that is, it encodes not only data but also the relational *context* of that data. Indeed, it smuggles this analog context into the digital realm. As database theorist C. J. Date argues, "It is fundamental to the hierarchical view that any given segment occurrence takes on its full significance only when seen in its context—indeed, no segment occurrence can exist without its superior."[17] For our purposes, such a database design wears its scalar structure on its sleeve and is thus accessible to analysis and critique.

The second type of database is the network form, which connects data as a series of interrelated nodes. This form, too, mimics real-world structures. A network database lists a set of entities (nodes) and a set of connections between them (edges). Representing data in this way has advantages and disadvantages: the network form emphasizes topological structure, which, as I have tried to show (particularly in chapters 3–5), is revealing of scalar relationships, or the ways that experiential potentials are disciplined into discrete (stabilized) scales with differential modes and constraints of access. Network structures imply not only scalar relationships but also *scaling* relationships—that is, they enable certain entities to scale along infrastructural gradients to occupy different regions of the scalar spectrum. It is no wonder, then, that the scaling functions of capital have greatly expanded under network-building infrastructure such as global shipping routes and protocols, digitization and tracking techniques, database management tools, and global trade agreements. But for all its representational and ontological efficacy, the network form also has drawbacks, such as a privileging of relations over things (edges over nodes) and the implication that all the nodes of a given network are ontologically and topographically similar, differing only in their network positionality and number of edges (that is, in their topologies). In Eugene Thacker's words, network representations "cannot account for the dynamics within networks; dynamics that show us a more complicated view of the separation between nodes and edges."[18]

For Codd, the great disadvantage of both tree and network database forms is that they overdetermine the structural relations of data with regard to encoding structure. Not only do all data domains have to be shoehorned into a tree or network structure, but that structure is encoded in the protocols of data entry itself. Data must be entered hierarchically and/or as a series of nodes and edges, and that structure is "baked in" to the database. This creates three dependencies that limit the flexibility of the database: ordering dependence, indexing dependence, and access path dependence. *Ordering dependence* is a requirement that data be entered and stored in the hierarchical order that the database models. In other words, the order is part of the relationship encoded in the data structure. This makes it difficult to append data later. *Indexing dependence* means that the database requires a separate index for easy (i.e., low-CPU usage) lookup when in actual use. *Access path dependence* means that algorithms written into software that access the database have to be coupled to the database's structure, and so "application programs developed to work with these systems tend to be logically impaired if the trees or networks are changed in structure."[19]

Codd's objections to these structures take the form of pragmatic complaints about the flexibility of the database structure as it affects both input interfaces and application-level access. A further, unstated implication of these objections is that databases should not be *complete* or *complex* models of the phenomena they represent. Instead, they should model small-scale relations without regard to larger-scale structures. While this may seem an abdication of control—a humble focus on modeling only micro relations— it is actually a way to format the lower-level structure of the data in a way that enables greater control over its symbolic manipulation. Certainly the representational grip on the world that such modeling approximates is not lessoned by limiting itself to simple relationships—we may remember Foucault's insistence that the most intractable and effective forms of control are those at the smallest possible scales, which he named the "micro-physics of power."[20] More prosaically, database theorist and formalizer David Maier tells us that "each row in the [relational database] table summarizes some object or relationship in the real world. Whether the corresponding entities in the real world actually possess the uniformity the relational model ascribes to them is a question that the user of the model must answer."[21] Hierarchical and network database structures maintain an analogical relation to the world outside the database, approximating the structural relationships they encode. They are less abstract and more context-rich. A hierarchical database contains an extra dimension to its encoded information, while a network database includes an entirely new ontological entity: the link. All of this complexity, from a database administration standpoint, hems the software in, forcing it to act like the world it references. Scale, for instance, becomes encoded in the database, not as data but as analogical structure.

The relational database, by contrast, purges itself of structural analogs, establishing instead true flat ontology, where every unit of information is, in its natural state, free of context or meaning. Relationships between encoded entities are now encoded as other entities on the same plane. As Date notes, relational databases "treat relationships as just another type of entity."[22] Ironically, then, relational databases demote relation from the status of structure to that of content. They trade analogical fit for mutational flexibility by generating a greater degree of uniformity among their data. The relational database thereby offloads the burden of representational fit to the "user" rather than the database creator. While all databases are technically digital, the relational database is digital in a more fundamental, ontological way: it scrubs its data of all analogical traces, rendering relationality completely flexible and mutable.

The practical effect of this shift is that the database becomes less structured, but the model of the entity or system that it encodes (e.g., the demographic constitution of a population, inventories of goods at various physical sites, a series of measurements, or the movements and signals generated by a digitally profiled person) becomes more dependent upon the structure of its associated access interface. Stated another way, the relational database grants less control to initial classificatory schema and input algorithms and more control to access algorithms. Or, more simply: *structure gets added later*, and there are almost no restrictions on the forms that structure can take. This is the (literal) power of the relational database: it frees data from any fidelity to the scalar relationships of the physical world.[23]

When representational control is distributed between database and algorithmic access, this creates privileged sites and modes of access, and also ensures that post facto control is always possible. This means that the representational control of a database is also always subject to reconstitution, and the "meaning" of a database subject to revision. As algorithm theorist Robert Kowalski emphasizes, logic structures determine the "meaning" of an algorithm, its domain applicability and essential nature, while control structures affect its problem-solving strategy, and thereby affect its behavior but not its meaning.[24] What is true of algorithms is true of databases: the logic component of a database is its classification structure, while its control component consists of methods of accessing its stored data. In hierarchical and network databases, both are retained to a greater degree, and power shifts away from the algorithm. This suggests that a database's "meaning" can be variably expressed or offloaded to access schemes. In this light, the development and continued preference for relational databases is not merely a matter of technical optimization: like most functional structural choices, it embeds cultural meaning and social relationships within its structural contingencies. Thus we can take Hayles's argument about the relational database's dependence on narrative one step further: the database form not only requires narrative post facto, as a form of interpretation, but actively assumes and anticipates it from the beginning. Whereas hierarchical, network, and object-oriented databases attempt to digitize narrative as context to be stored internally, the relational database initially de-narrativizes its data while structuring it, but then requires a narrative structure (in the form of the query) to be in place before its data can even be accessed. In other words, the primary role of narrative vis-à-vis the relational database is not to explain its results but to potentiate and structure the sorts of results it can produce in the first place.

This is a scalar shift: the relational database encodes difference at the smallest possible scale, generating uniform, standard modules maximally susceptible to algorithmic manipulation that can in turn enable large scaling operations. The relational database gives up its ability to encode scale so that it may gain greater scaling potential when reconstituted from the outside. In fact, it is precisely because the inherent narrative elements of network topologies and hierarchical dynamics are purged in relational database encodings that their virtual narrative dimension becomes so enlarged. The database as back end is reduced to scalar homogeneity, while the front end—that is, the speculative aggregation enacted by the query—is endowed with the power of the cosmic zoom.

QUERY AND THE GENERIC SCALAR EVENT

A database's structural relationship between data storage and data retrieval—its interface—is the key relationship that effects the distribution of power or control of its contents. In this section I explore the nature of the database interface as an active and open-ended system for the production of new aggregates. This is the primary means by which databases mediate scale.

The formal, low-level process of accessing a populated database is the *query*. A query is, as the name would suggest, a question. In a hierarchical, network, or object database, one asks for data and receives back data plus context. As I discussed previously, the two are inextricably linked. We can think of hierarchical relationships, then, as an implicit component of any question put to hierarchical database structures. Relational databases, on the other hand, can only be queried about sets. That is, when one puts a question to a relational database, it is in this highly abstracted form: which objects satisfy the condition of belonging to a set that includes properties $p_1 \ldots p_n$? The "answer" to such a question is a subset of the database's stored information, its structure determined by the form of the query. Each query, then, *produces* a newly structured set of data that is novel in the sense that the relationships it expresses or models were not present in the initial act of entry into the database. In Maier's characterization, a relational database models real-world properties as "tuples" that are entered into its folds, but only in a minimal, flattened way. The real magic occurs later, in the query: "A *query* is a computation upon relations that yields other relations."[25] Any relational database schema (such as the currently dominant MYSQL) includes an entire "query language" with which to formulate these complex structures. Its various commands (actually algebraic or calculus expressions) each perform some algorithmic operation on a subset of the data found within a particular "table" of a database and thereby transform that data into a new structure bearing new internal relationships. Ontologically, then, a relational database query *produces a new assemblage* of entities with a novel structure found neither in the object relationships initially sampled upon data entry, nor "in" the populated database prior to its being queried.

If, as I suggested earlier, the relational database enacts a strategic shifting of control from encoding structure to interface or query operations, and query operations are ontogenetic, then (relational) database-driven scalar mediation invests the interface with the power not only to stabilize new scales but to populate those scales with new entities. Of course, these new-found digital powers are neither necessarily creative nor necessarily liberatory. As Alexander Galloway reminds us, "An interface is not a thing; an interface is an effect." It is "a general technique of mediation evident at all levels; indeed it facilitates the way of thinking that tends to pitch things in terms of "levels" or "layers" in the first place."[26] Interfaces, as surfaces, promise access to layers beneath, but actually they produce those layers; they make the cuts that stabilize the domains they supposedly bridge as windows or thresholds. For Galloway, mediation as a regime of significa-tion and control ontologizes interfaces as particular sites of access as part of a more general sorting process that tends to stabilize forces as things. As I have noted throughout this book, scalar mediation does exactly this: it produces, as a medial effect, stabilized scalar domains that it then oper-ates upon in acts of resolution. The trans-scalar interface is produced qua threshold and yet stages a negotiation of difference (the trans-scalar en-counter) that reveals ontological difference in the form of ecological detail. The trans-scalar interface is thus a medial effect in Galloway's sense, but also a site of the agential cut in Karen Barad's sense (as elaborated in chap-ters 1 and 2). With this dual sense of "interface," then, let us look at one in action.

Cary and Michael Huang exploit the scalar database model to full effect in their Flash-based, interactive version of the cosmic zoom. Beginning in 2010, they produced what would become, two years later, *Scale of the Universe* 2, combining the infinite zoom of the Eames films with the curi-osity cabinet model of the 1982 *Powers of Ten* book (figure 29).[27] When the user manipulates the slider at the bottom of the interface (there is also a linear, cinematic play-through mode), the screen "zooms" up or down the scalar spectrum. The continuous scalar plane is not, however, isomorphi-cally matched with a contiguous spatial one; rather, the "space" we traverse is abstract. Particular objects are positioned according to their relative size on a monochrome surface that resolves new scales in response to the user-

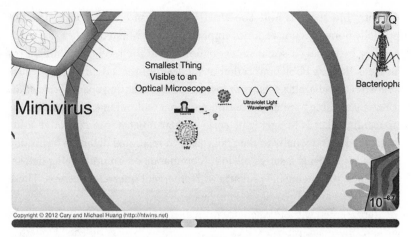

Figure 29 ▸ *The Scale of the Universe 2* by Cary and Michael Huang. http://htwins.net/scale2/.

controlled zoom. The effect is to aggregate entities of completely different classes and real-life contexts. We find side by side, for instance, the width of a silk fiber, an infrared wavelength, a mist droplet, and a white blood cell (at 10^{-5}). Also found together are the Homunculus Nebula, the Kuiper Belt, and the current distance from Voyager 1 to earth (all at 10^{13}).

In this devious remediation, scalar boundaries are clearly marked, in classic cosmic-zoom form, as both labeled powers of ten and quadratlike spatial boundaries (here, the circles from *Cosmic Voyage* rather than the squares of earlier instantiations). The visual smoothness of the zoom is enhanced through its vector-graphic implementation. Vector graphics are produced based on mathematical description rather than encoded pixels (as in bitmapped graphics). While vector images can only be displayed *as* pixels, their code contains no "baked-in" pixel parameters. This function— the translation of mathematically defined shapes and vectors as displayed (pixellated) patterns—is offloaded to a vector renderer, built into application code, web browser code, or in this case, Adobe Flash's rendering engine. Vector images are unique in that they *have no native scale*. No scalar coordinates are encoded in their file format. Only when a vector renderer calculates a vector's set of trajectories in relation to a grid of pixels is a scalar relationship stabilized. *Scale of the Universe 2* takes advantage of this translational system to produce, whenever the user moves the scale slider, a smooth, continual renegotiation of the relationship between fixed resolution of the pixellated screen and the displayed scale of the images. This work implements its virtuosic scaling interface only to disarticulate the zoom from spatial coordinates—the scalar spectrum is continuous, even smooth, but space is not. The visual aggregation of scale-proximate but spatially noncontiguous entities emphasizes rather than occludes scalar difference by forcing a scalar renegotiation not at the level of the milieu (as in Kees Boeke's book) but rather at the level of the assemblage. Because it abstracts the scalar spectrum instead of representing space, *Scale of the Universe 2* is able to retain a continuous zoom without invoking contiguous space. Its effect is to aggregate scale-similar objects into collective units organized orthogonally to the scalar axis. In relational database terms, the interface performs a series of "join" commands on an underlying dataset of entities, producing new subsets with emergent interrelationships. These new subsets are intrascalar assemblages.

Represented without reference to an environment or medium that would ground these assemblages, they are certainly reified as objects. Nonetheless, a disciplinary scrambling is in effect here, a playful shuffling of discrete forms into scalar categories that retain and even highlight their differences in structure as well as their distribution in space and positional coordinates within our disciplinary knowledge structures. When two unlike objects appear together, our attempt as users to connect them requires traversing their distinct categorical orders, a virtual journey through disciplines to arrive at common ground, a game of "six degrees of scale." How to process the relations between a silk fiber, a blood cell, and a wavelength? The Huang brothers here accomplish something similar to the absurd lists of dissimilar objects that Michel Foucault credits Jorge Luis Borges as deploying against disciplined knowledge structures: "He does away with the *site*, the mute ground upon which it is possible for entities to be juxtaposed."[28] *Scale of the Universe 2* replaces environmental space with a monochrome background, an abstract mist or plane of consistency that dissolves the ordered grid of knowledge Foucault calls "a *tabula*, that enables thought to operate upon the entities of our world, to put them in order, to divide them into classes, to group them according to names that designate their similarities and their differences."[29]

For Foucault, language produces the order of the *tabula* but is also capable of scrambling it, dissolving it. Something similar could be said of scale. *Scale of the Universe 2* reorganizes and aggregates objects into assemblages according to the logics of new virtual spaces by deploying scalar continuity as an organizing principle, disrupting spatial contiguity and categorical positionality. By scrambling the grid of predefined object relationships, it catalyzes a new, creative chain of speculative knowledge production, prompting us to ask, starting from a disciplinary *tabula rasa*, what makes these objects in uncanny scalar proximity similar to and different from one another.

Items entered into a database are minimally encoded. As Geoffrey Bowker and Susan Star argue, any classification scheme acts as a funnel during the encoding process, forcing all entries to be expressed as one of the available categories, and thus captures only a small subset of potential properties.[30] In the case of scale, however, such reductive or thin para-materization is inversely related to ontogenetic vibrancy. This is because *scale inheres in the virtual space of becoming between objects.*

Any set of things can contain objects of different sizes. Size is a thin parameter: a reductive, unidimensional application of an arbitrary but un-ambiguous unit of measurement to the extension of a thing. It is among our most basic classifications, and one of the most common relations encoded in databases. Scale, however, emerges between two or more objects as a spatiotemporal domain that encompasses and enables the expression of relationality. This is the ecological concept of scale that I have argued for throughout this book. Properly understood, scale in this sense is emergent, better expressed as a question than as an answer.

We can never fully exhaust the ontogenetic potential of scalar difference, because we cannot know in advance the full set of potential becomings that exist within any set of entities. The question of scale is the question of what things can become, how they can be affected. It is a variant of Spinoza's famous provocation, "No one has yet determined what the body can do."[31] Where are the potential articulations within a body that express (divide into) two different objects? What is the third body (assemblage) that any two objects can form? These are scalar transformations that begin with a gap between things. Scale is a mystery inaugurated by the most banal form of collecting: $x + y + \ldots$?

A relational database is like a child who scoops up a fistful of this and a fistful of that. Query: what have you got in there? Command: Open your hands right now! The nursery floor, littered with dropped objects, may be a dangerous terrain of potential injuries and breakages, but it is also an emergent scalar milieu.

Throughout this book I have examined the process by which scalar re-lationships are stabilized through processes of mediation that enable trans-scalar encounters. *Scale of the Universe 2*, and the database form more gen-erally, prompt us to ask about a special case: what happens when we collect and aggregate objects of the same size? Does this banal case belong out-side of questions of scale? The Huang brothers suggest otherwise; their underlying database has classified objects by size, while their cosmic-zoom interface, a form of query, aggregates them within spatiotemporally defined and delineated scalar domains. I would suggest that the juxtapositions presented by this interface prompt a dual question: How might these objects interact? and What do all of these objects have in common? Scale frames both questions: potential interactions are produced on a plane of encounter, a scalar domain of contiguous space. Encounters *across* scales are mediated in the special ways I've enumerated in this book, but encounters *within* scales are another question entirely.

Potential interactions within scales are events. Scalar domains are defined by the events that fall within their aegis. As I argued in chapter 1, however, scalar domains rely for their stabilization not on singular events but on *typical* events. At 10^{-7} meters viruses infect tissues, ultraviolet wavelengths of light are propagated, and transistors amplify and switch electrical signals. Technological mediation is necessary to fully trace the implications of these activities, which cross scalar thresholds to produce their typical effects: a colony of viruses can only be resolved at the scale of a cell or an organ, and its larger effects are only visible at the scale of its infected host. Similarly, ultraviolet wavelengths gain meaning only when they are received at the scale of optical receptors, the nervous system, and so on. Transistors perform their individual operations at the this scale, but their most dramatic effects are only resolved when a television flickers to life at 10^0 or a computer screen displays a Flash animation. These are different interactions and different assemblages, however, than belong to 10^{-7} meters. What would it mean to trace not these entities' individual transscalar becomings but rather the collective becomings they form within their single scalar domain?

The transistor, the ultraviolet wavelength, and the HIV virus. What speculative ecology is this? What do they do together? This is in the first instance a database question, a query. We are given multiple objects that belong to the same class, that possess the same size attributes, but that are not ordinarily encountered together. It is a special case of a 1:1 scalar relationship. The question, more properly, is this: if the database aggregates these objects based upon the property of size, what temporal domain unifies them?

A scalar domain is composed of contiguous space and continuous time. Where are its boundaries? The classic cosmic zoom extends the first indefinitely while varying the second (or rather, relativizing it to the first; this is, diagrammatically, the operation of scalar collapse). The database-driven zoom aggregates objects along both an axis of scalar difference and an axis of aggregation. The first axis organizes trans-scalar *encounters*, the second intrascalar *events*. Crucially, however, these events are generic. The HIV virus replicates itself in space and time. The transistor alters the current of an electrical signal, changing its state. Ultraviolet waves oscillate and bounce. It's just a typical, if harrowing, stroll through the neighborhood of 10^{-7} meters. What kind of events are resolvable on a regular basis? No electron energy level jumps, just glycoprotein adsorption. No computation, just individual signal switching. No Jim Morrison posters, just "black light."

This is what I mean when I posit that scales are generic domains. A scale isn't defined by gridded boundaries in space or time, despite the fact that we can stabilize scales from the top down through the application of such grids (quadrats). A scale is the abstract space and time in which a set of typical events take place. Even 10^{-7} meters is as big as the universe, but only certain events take place there. This isn't to say that they can be exhaustively enumerated; the potentials for intrascalar events are as vast as those of trans-scalar encounter. Database aesthetics can help us to define and understand scalar milieus as horizons for generic events—not *despite* their categorical flattening but *because* of it: the more reductive the initiation into a coded matrix of relations, the greater the potential for emergent complexity on the axes of both intrascalar aggregation and trans-scalar encounter. Provided, of course, that we ask the right questions.

DATABASE MILIEU: NEOLIBERALISM AND SCALAR RECURSION

If database media render visible both the markers and the potentials of scale as generic event, inaugurating nearly limitless virtual aggregations (and thus new virtual milieus), they also accelerate the dissolution of the human subject. This will be the focus of the final three sections of this chapter.

Caught like a fly in a web spun of gossamer database tables, the human subject is in the process of being reconstituted by ubiquitous digital surveillance, globalized mechanisms of finance capital, and a trans-scalar computational infrastructure that, as Mark Hansen notes, is "only indirectly correlated to human modes of experience."[32] Database, then, an ostensibly human invention intended to render visible new scalar milieus, and thus inaugurate new scalar events, has turned out to be monstrous in the classical sense: it has created a world that is not recognizably human in that it no longer addresses the human and no longer respects the boundaries of the liberal subject. As John Cheney-Lippold argues, referencing Google's algorithmic assignation of "female" as his gender,: "Google's misrecognition of my gender and age isn't an error. It's a reconfiguration, a freshly minted algorithmic truth that cares little about being authentic but cares a lot about being an effective metric for classification. In this world, there is no fidelity to notions of our individual history and self-assessment."[33] The database, it seems, has taken identity, which we traditionally believe to be a form of autonomous self-expression, out of our hands. Seb Franklin also emphasizes the externalization of experience incumbent to the digital mediascape, but in the context of scalar—rather than merely algorithmic—abstraction:

> What is this process [of digitization] if not a restaging of the conversion of human time into individual units on a single, abstract scale, which can then be used to work out an equivalence between qualitatively different objects (wheat, steel, etc.), but at a more universal level? This dream of universal digitalization and valorization is the dream of capital, and the logical system through which it proceeds is that of control.[34]

Here Franklin reads digitality as a technique of abstraction and universal equivalence. This argument parallels and updates Marx's analysis of money as a particularly destructive form of abstraction and universal equivalence. While for Marx the scale of capital is implicitly only a derivative effect of this abstraction, for Franklin scale becomes the primary output. For those following the analogy, money has become code and capital has become abstract scale. In the vocabulary I have been developing here, we could say that database-driven media collapse scale and thereby compose new virtual milieus (the evental genres I discussed in the previous section) as substrates for further scalar differentiation. Franklin's concern, however, is that the abstraction involved in this process—the relational database, which, as we've seen, leverages abstraction as a compositional force—has captured the human subject in an inexorable and deeply alienating flow of composition and decomposition that blithely ignores and subverts individual (as well as collective) agency, just as for Cheney-Lippold it produces and assigns identity markers that may dramatically diverge from our own self-narratives. I take it for granted, then, that our contemporary milieu, driven by computational capital and mediated by the database form, is deeply alienating, in the sense of distancing our perceptual lifeworld from our narrative sense of identity.

What I wish to emphasize here is twofold: first, that this process of alienation is scalar in nature and, second, that it is a red herring. To critique capital from the perspective of the threatened (human and humanist) individual is to miss its most pernicious effects. Late capital has become a trans-scalar force, and not accidentally; rather, that has become its central characteristic and the source of both its resilience and its destructive power. Tracing the scales of capital has become imperative, and continued fixation on the presumptively scale-fragile human subject merely plays into its hands by chaining us to one scale while capital plays out its effects at others. Thus, while the database form abstracts scale, it does not, I will continue to argue, collapse it.

Late capital depends upon innovative, network-enabled forms of scaling, but relies just as heavily on scalar differentiation to maintain differential gradients between sites of exchange, centers of accumulation, and striated milieus of extraction. Concomitantly, capital buttresses the human subject—even as it alienates it. There is no contradiction here for capital. Wendy Chun makes a similar point when she suggests that networked media addresses "not a 'we' but always a 'you.'" "Instead of depending on mass communal activities . . . networks rely on asynchronous yet pressing actions to create interconnected users. In network time, things flow non-continuously."[35] In other words, capital, or rather its digital medial expressions, are individuating. Chun's notion that digital media—she has called it "n(you) media"—functions at the behest of capital through modes of individuated, asynchronous address is a key insight.

More than a mere marketing ploy, asynchronous, individuated address is a technique of containment and control, a continually renewed (Chun would say, continually updated) scalar composition that limits the individual's trans-scalar perspective and potential, reserving for capital the fecundity of database composition. I can access previously unimaginable scales of data, virtually travel the world, visualize alien scalar milieus from the nanoscale to the galactic. I daily participate in massive social networks, and yet I am constantly reminded that my identity consists of a punctiform scalar perspective, that I am defined by a continuous and coherent narrative of which I seem to be the author—that is, that I inhabit one and only one scale. Far from threatening the classic liberal human subject, database capital's techniques of address absolutely depend upon this tired narrative. To explore this scalar paradox of database subjectivity, I'll turn for a moment to perhaps the weirdest cosmic zoom discussed in this volume thus far.

The opening sequence of Neil Burger's 2011 film *Limitless*, sees Eddie Morra (played by Bradley Cooper) standing on the edge of a high-rise roof in Manhattan, about to jump to his death to avoid capture by a crew of goons who have just burst into his penthouse apartment. "I'd come this close to having an impact on the world," he narrates. "Now, the only thing I would have an impact on is the sidewalk." The camera, ostensibly representing his point of view, plummets over the edge, but just as it would make contact with a taxicab on the street below, it begins to zoom laterally instead of vertically, at street level. The camera continues to zoom, uninterrupted, for the next minute and fifteen seconds. Clearly calibrated to emulate classic cosmic zooms such as *Powers of Ten*, the sequence inverts their locomotive signification. While *Powers of Ten* is a long zoom that signifies itself as perspectival movement, in *Limitless* we traverse large sections of Manhattan but never feel we're moving at all. This is because the effect of the zoom is not hidden but enhanced. The image feels flat, two-dimensional, even though it teems with movement: cars driving, pedestrians walking, lights and signs flashing. We constantly sense that we are running out of resolution, even as new detail emerges. The periphery of the image is indeed always getting fuzzier. Objects don't pass us by so much as exit the edges of the frame. This is to say, no object in the frame alters its relationship with any other object over time—no objects become proportionally larger as the camera nears them, as Eva Szasz visually acknowledged with her two-dimensional moon in *Cosmic Zoom*. In other words, the camera never gets closer to any of the objects it ostensibly films, even as we zoom closer and closer (figure 30).

Eventually the shot reaches Times Square and zooms into a jumbotron displaying—recursively—the Manhattan cityscape. This quickly resolves itself into orange pixels that become an MRI scan of a human head. As the camera zooms closer, into the brain, individual neurons are resolved. Inside the axon of one neuron, however, we resolve another entire brain, with more neurons, one of which contains another brain . . . until that final brain resolves into a satellite view of all of Manhattan. Finally, the camera zooms into a single rooftop, signaling the end of the sequence, as well as the end of the opening titles, pretty much right back where we started, spatially and narratively.

A network graph (figure 31, left side) of the *Limitless* zoom reveals a thin multiscalar ecology. It spans a paltry eleven orders of magnitude and emphasizes a central feature that is quite visible on-screen: the human brain. Unlike previous examples in this book, this network graph fails to reveal features of a trans-scalar constellation not readily apparent in a cursory viewing of the sequence. This is because the network graph form fails to capture the sequence's most unique feature: its recursion. Just as the topological structure of linear cosmic zooms is difficult to consciously visualize in time-based viewing, so are the recursive topologies of nonlinear zooms. The following diagram (figure 31, right side) abstracts this sequence in a different manner, capturing the relationship between temporal progression (lines emerging under the square starting scale of the human body) and

Figure 30 ▶ Infinite zoom of banality. Frame from *Limitless* (2011).

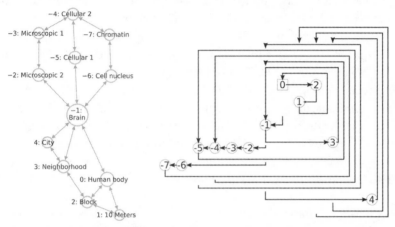

Figure 31 ▶ *Limitless* (2011) mapped as a scalar network graph (*left*) and recursive diagram (*right*). Scales are labeled as powers of 10 meters. Diagrams by the author.

recursive return (the portion of lines that rise above the starting scale). In this diagram, stacked arrows above a particular scale indicate moments of return, or scalar repetition. Unlike a network graph, which more readily demonstrates the connectedness and topological centrality of particular scales in a multiscalar milieu, this graph preserves the unfolding of medial time—you can follow the sequence as a radially expanding progression if you read top-to-bottom and left or right in the direction of the arrows. The recursive structure that it renders spatially, however, provides a map to the film's exploration of the scale-recursive milieu of database capital.

Tracing the circuits of this zoom reveals a frustrating maze, where every promise of escape to another scale leads back to familiar ones. Each return from alterior scales seems to enhance the familiar's trans-scalar potential for encounter: the potentials of the human brain are, in the film's narrative, unlocked through the molecular scale of pharmaceuticals and the macro scale of global finance capital, each enabling access to the other, but only through an insistently repeated re-re-reactivation of the individual. Yet, even while each recursive loop contains different potentials than the last, the overall milieu is ultimately in a steady state, and there is no escaping the maze.

One of this zoom's most striking features, in the context of its emulation of *Powers of Ten*, is its deliberate banality. There are no spectacular galaxies, no awe-inspiring events. Just life on the street. We travel from Eighth Avenue through Harlem and then to Times Square, yet we don't feel we've moved at all, as semiotically the zoom negates movement (as I discussed in chapter 3). New York City, in this sequence, becomes ceaselessly animated but geographically inert—evoking a kind of metropolis-scale Brownian motion. The differences exhibited by this cosmic zoom have more to do with partitioning space than with transported perspective. As Benjamin Bratton argues in *The Stack*, his monograph on the interrelated scales of planetary computation, "The *City* layer's basic building unit is the interfacial partition, the physical or virtual boundary that is made systematically reversible by the situation of its functions within larger computational systems."[36] In the *Limitless* zoom, the camera zooms through numerous vehicles, which serve to partition the otherwise homogeneous space between interior and exterior, private and public. As Bratton notes:

> Inside cities, the line, the gate, and the wall separate inside from outside, quasi-private from quasi-public, and generate and direct flows accordingly . . . their ability to interiorize and exteriorize is also always reversible, and so even when it does not move, which side is being exteriorized by the other can switch in the blink of an eye.[37]

The operative model of control and order here is not geography or even architecture, but circuitry: the switch. As figure 31 visualizes, scalar recursion produces a topography that affords nearly limitless switch points and reversals; unlike the rhizomatic scalar topology I explored in the previous chapter, recursive topology privileges the switchback, a diversion of scaling processes back into previous scales. Here, particular scales engineered to be well connected offer less potential for embarkation than they do as likely return destinations. In the world of database capital, all roads lead *back* to Rome.

Limitless's play of exteriorization and interiorization as switched state evokes, for me, the famous dolly-zoom from *Vertigo*, where Alfred Hitchcock simultaneously moves the camera forward, closer to the subject, and zooms out, matching the two rates of change such that the field of view remains constant but the relation between foreground and background rapidly and disconcertingly shifts. Space appears to elongate as the background both recedes (in distance) and becomes better resolved. The effect is meant to mimic Jimmy Stewart's subjective experience of navigating San Francisco's vertiginous urban infrastructure, an experience characterized as the film's titular malady. While *Limitless* employs a different sort of zooming technique, its effect is similar: it tracks the switching nodes of urban infrastructure not as navigable spaces experienced by a perambulator, but as a seemingly unbroken contiguity of homogeneous signal structures—streetlights, stoplights, motor neurons—subdivided only by membranes delineating the private from the public. The opening shot zooms alternately through the interior spaces of cars and the human skull and the exterior spaces of streets and billboards. A later zoom in the film passes alternately through streets and interior rooms, cataloging locations and actions during one night in Eddie's life. Here the zoom is explicitly divorced from the perceptual flow that characterizes individual human inhabitation and is instead calibrated to the trans-scalar processes that orient and subject the individual in global technoculture, optimizing it for these dynamics.

This optimization takes place at multiple scales, from chemically conditioned responses in the neurons of the brain to public infrastructure (and its many codes of communication) to the even larger structures of the global financial system and its constitutive networked power structures. In the film's narrative, Eddie is a struggling writer at the periphery of the forces concentrated in the city—including, of course, the publishing world that subjects him to (unmet) deadlines but cannot keep him financially solvent. He manages to obtain an experimental drug called NZT, which allows him to access "one hundred percent" of his brain. NZT is the switch that brings him inside the trans-scalar structures of capital that had previously ordered his life from the outside. Suddenly—*click*—he discovers capacities he never had before. He can now access any and all information he has ever absorbed, even if it is beyond the grasp of his conscious memory. Numerous scenes depict him assembling fragments of data (snippets of legal code, biographies, topographical features of his surroundings, etc.) into a pattern that reveals potential action in the present moment. He completes his novel in one night, recalls obscure facts to put together a legal case for his neighbor, repeatedly evades capture by unsavory characters, and of course becomes a stockbroker. Because he can absorb nearly limitless quantities of information and bring his entire experiential archive to bear on any problem he encounters, he quickly gains privileged access to patterns and structures that far outscale the human individual. The predictable result is that he comes to understand the large-scale workings of finance capital and can easily exploit them to gain wealth and power.

Paradoxically, Eddie in *Limitless* serves as both the perfect neoliberal subject, willing to be reshaped into a sharp tool of capital, a libidinal concentration of its networked, global energies, and a scalar fantasy of regained control over a multiscalar system far removed from the human lifeworld. The zoom, then, not only figures the scalar ladder of neoliberal technoculture—with its compulsory self-shaping into a tool of access, with the goal of being switched from the outside to the inside of whatever structure one encounters—but also, like every form of the cosmic zoom, presents the fantasy of optic integration across scales, a potential form of mastery through networking. This is the computational version of what Deleuze and Guattari dub "desiring machines," which produce subjects, and everything else, through the principle of connective synthesis.[38] In neoliberal technoculture, desiring machines immanently produce not only the human subject caught in its folds, but the folds themselves, composed of heterogeneous materials self-organized into homogeneous trans-scalar connective circuits. The *Limitless* zoom visualizes these circuits as infrastructure. Eddie, as neoliberal subject, gains the ability to traverse the codes produced by capital through a becoming-exterior.[39] That is, to become an insider he must become exterior to himself, a process that takes the form of a de-linearization: the world as dictated by experience becomes flattened into an abstract, synchronic plane, with all of its details, its gleaned information, its data, ready at hand. By de-linearizing his narrative trajectory, granting him privileged access to his total experiential field, and thus to structures outside of the narrated self, NZT *turns Eddie into a database.*

Perhaps more accurately, Eddie, in becoming-database, realizes that he always was one, *in potentia.* He learns the query language. What was required, and what the McGuffin of NZT affords, is a recomposition of the codes and accretions that are/were Eddie Morra. Like the film's opening zoom, this is a recursive realization: NZT grants Eddie access to himself from the exterior, from outside his linear, subjective narrative. In viewing himself from the outside he is recomposed into a different sort of subject (and object): one that has gained the capacity of recomposition. This is a form of alienation that always leads to something new, even as it mines the self to grant access to the outside. The *Limitless* zoom evokes the freescaling aesthetics and optic aggregation of disciplined knowledge of the classic cosmic zoom, rendering it recursive and recontextualizing it in the banal milieu of finance capital (figure 32).

Figure 32 ▸ A brain inside a neuron inside a brain. Frame from *Limitless* (2011).

What new form of agency is this? The *Limitless* zoom captures the perspective of being inside a database, being confused and overwhelmed by the forces that touch your life but exist in scalar milieus to which you have no access. As I argued in the previous section, as a linear subject you can coincide with evental milieus outside of your own (by occupying the same space) and yet remain unable to participate in their dynamics, failing to catalyze events at alterior scales. Eddie, "this close to affecting the world," holds out the promise of trans-scalar access amid the banal landscape of late capital, the traversal of which takes the form of the endless zoom rather than true mobility. Eddie's promise of becoming-database, however, is figured as the code switch between endless lateral zooming through urban infrastructure and the (now far less common) scale-jumping of the classic cosmic zoom. These classic jumps are now associated with the database form, and thus have taken on a recursive structure. One queries what one has in order to compose something new. One can access new scales in these encounters, but only through recursive abstraction. The director of the film, Neil Burger, has referred to his invented technique as a "fractal zoom." Just as Benoit Mandelbrot characterized "fractals," a class of particular non-Euclidean geometric shapes that repeat their structures at every scale, we discover by watching this sequence that following capital's patterns, gaining access to its trans-scalar secrets, means traversing a new form of scalar collapse that is not a fleeing of the self—Scipio's out-of-body experience as an escape to other scales—but rather its rediscovery from outside, again and again and again: the self shot through with capital, capital shot through with the self.

To compose this sequence, Berger shot footage at a series of intersections with three cameras on a stationary tripod filming the same subject with three lenses of different focal lengths. The postproduction CGI team, led by Tim Carras, composited the shots using techniques similar to those used in the Eames films, based on a series of image plates. The primary difference in *Limitless* is that the image plates are moving images rather than still photographs. They would digitally zoom into the footage from the widest lens "until we started to run out of resolution," at which point they would shift to the medium lens, then the longest lens, then widest lens at the next location, and so on. Carras perfectly describes what I called "arcs of resolution" in chapter 3. And as I have argued, the "fractal zoom" doesn't actually go anywhere, instead collapsing upon itself, heralding not just database-driven scalar mediation but a simulation of what we might call *recursive database subjectivity*.

Carras evokes the powers of ten when he describes the virtual zoom lens he created algorithmically: "If you work out the mathematics of it, which we had to do inside our software to get it to work, it's essentially a zoom lens that goes from a 25mm to a 10-to-the-36th-power millimeter."[40] This figure is not merely an equivalent or analogous degree of zoom, but a real simulated model of such a lens in software. As Carras explains:

> We wrote an expression for the focal length of the camera that was based on an exponential function so that the speed would be a constant for the entire sequence and the focal length would just double say every 28 frames. From there we had a camera that would reliably stay at the same speed . . . once you get past the first three or four hundred frames, you're dealing with such small values. You're essentially zoomed in on what would have been the size of one pixel. If you move a plate by say a thousandth of a pixel, it's the equivalent of jumping half way across your frame, because the camera's zoomed in so tightly by then.[41]

Carras and his team of twenty visual-effects artists worked for two months on the sequence, modeling a trans-scalar zoom lens that produced a single shot in ways that pushed even the boundaries of their high-end compositing software (Nuke). This is the virtual equivalent of the Eameses' animation stand, if they had used one with a track hundreds of feet long that kept pushing in throughout the entire film, as they constantly swapped out image plates along the way. Why not just reset the (simulated) apparatus for each new plate, as the Eameses did? While the viewer will ordinarily not have access to knowledge of this compositional technique, ontologically it functions as a model for recomposing time and space across scales into a single evental milieu rather than taking the scenic route of jumping to new perspectives within a scale-variegated space. This, then, is the scalar privilege—and the milieu—of capital.[42]

Eddie Morra, as database, as postdigital subject *par excellence,* has a perfect memory. He stores everything and can recall it instantly. Yet while memory, as traditionally conceived, is oriented toward the past, the database is oriented toward the future. Memory is an encoding of an event into a context, while database is an encoding of an event into virtual potentials for future compositions. Memory, when activated, continuously rearticulates links between past states and the individual subject as experiential continuity. The database, on the other hand, expresses no fidelity to a continuous, molar subject; rather, it composes new assemblages out of old data, an operation that quickly exceeds the scalar boundaries of the individual. As a model of mediated human subjectivity, then, the database form marks a profound shift in temporality and scale that punctures the carefully guarded boundaries of the human. Wendy Chun has noted a modal shift, as digital media reaches critical mass, from "memory" to "storage." (Biological) memory produces and arises from the classical self, from a unified perspective, in a narrative mode. "Memory contains within it the act of repetition: it is an act of commemoration—a process of recollecting or remembering."[43] We might say that memory possesses, or is produced within, continuity: both with itself (as narrative flow) and with the nodal subject to which it belongs and always maintains contact. In this sense, memory has a particular scale. (Computer) memory, by contrast, functions in the mode of storage:

> Stores look forward toward a future: we put something in storage in order to use it again; we buy things in stores in order to use them. By bringing memory and storage together, we bring together the past and the future; we also bring together the machinic and the biological into what we might call the archive.[44]

Let us consider the scales of storage. Rather than taking an iterative form in contact with a particular subjective position, a core perspectival scale, memory as storage is recursive: it always extends outside of itself, protocologically, producing a new collection, a new perspective from which to make the same call again. It does not return to its iterative home but travels onward, infinitely. Limitless. This aggregative traversal of a hoarder that becomes the horde that becomes a hoarder continually perforates scalar boundaries in both space and time. The relational database is, as David Maier reminds us, "memoryless." Unlike analogical databases, its particular states "will depend only on the current state of the relation and not on its history of previous states."[45] In this sense it is like Philip Morrison's atom, explained in chapter 4. The relational database represents a world made anew with every query: an accretion of the past reconfigured, through a radical break, into an always nonlinear future. An entity's degrees of freedom within a digitally mediated structure are therefore a function of the state of that structure and not the previous conditions of (analog) relationality that it encodes. This truth is felt, in ironic form, by many of us when confronted with ubiquitous surveillance as a defining characteristic of our current milieu.

We face, as surveilled digital subjects, a strange dissociation between the scale of our narrative histories (encoded in narrative memory) and the scale of our database histories (encoded in computer memory). This dissociation becomes more consequential as databases grow in scope and detail through massified inputs (surveillance of various sorts) and ever more complexi-fied outputs (linked to various forms of power and transformation). As far as the technocultural zeitgeist is concerned, popular engagement with both of these poles of database identity seems to have reached something of a critical mass with Edward Snowden's release, in 2013, of an enormous cache of data pulled from US government servers that revealed the scale of the surveillance programs run by the United States and other govern-ments around the globe. It is precisely this scale, the database version of so many identities, all massed together, along with the scale of the apparatuses of digitization, that captured the public's imagination and incited its anger. Of course, the incident's viability as a popular story was in part due to the way it played as an old-fashioned narrative: Edward Snowden (whistle-blowing superspy or treasonous turncoat?) smuggles the biggest database of secrets in espionage history out from under the US government's nose, escapes to Hong Kong (then Russia), a lone individual outsmarting and outmaneuvering the most powerful structure on the planet. This narrative, a form of linear memory, thus features a radically scalar asymmetry that is structurally inverse from the other narrative, the database narrative that his leak confirmed: that each of us is in fact subject to government surveillance that reconstructs our identities within a massive database somewhere, sub-jecting us to the control of the security state. The fact of this inverted scalar asymmetry functioning within the primary scalar asymmetry inherent in the database versus the individual history makes the Snowden saga ripe for scalar representation. Cue the cosmic zoom.

While the intertwined stories of Snowden's spycraft and the mass sur-veillance spycraft he revealed are most effectively narrated (in the midst of their unfolding) by Laura Poitras in her remarkable documentary *Citizenfour* (2014), it took Oliver Stone's visual flair and Hollywood budget to render these antinomies into the medial form of the cosmic zoom. In *Snowden* (2016), Stone gives us, at almost exactly the halfway point in the film, a dramatic cosmic zoom that serves to illustrate both the scale and database logic of US government surveillance. The scene begins with Snowden (played by Joseph Gordon-Levitt) sitting at his workstation, busy at his daily grind. Narrating over the scene, he explains that his routine involved SIGINT, or signals intelligence.

The zoom begins as he gives an example of how networks are followed and created through routine surveillance. "Let's say your target is a shady Iranian banker operating out of Beirut." A bright blue, flickering series of lines leap out of one of the names on his computer screen. As we zoom into this stylized flow of data, we see various video streams playing within it, depicting the Iranian banker. More video streams fly by, continuing down the blue stream. The camera follows this as it breaks apart into many more streams, radiating out from a map of Lebanon. Snowden explains that surveillance targets are chosen from the contacts of the original target. We follow one of these streams to several sites in the United States, until it explodes into an enormous global network, populated by hundreds or thousands of portraits and names. "Three hops from anyone with, say, forty contacts . . . you're looking a list of 2.5 million people." The image, still following some of these data streams, superimposes a series of close-ups on faces: first Snowden, then a series of ordinary, unrelated people in everyday situations. Snowden: "And there's that moment when you're sitting there and the scale of it hits you. The NSA is really tracking every cell phone in the world. No matter who you are, every day of your life you're sitting in a database just ready to be looked at." The images resolve into a software Privacy Policy agreement, which an unseen user clicks. The screen decomposes into many lines of a stylized data stream. The camera pans with the stream, but then zooms back and rises up to reveal thousands of such streams, all flowing into an enormous black globe. "Not just terrorists, or countries, or corporations," intones Snowden, "but *you*." As the camera zooms further back, the giant globe and the thousands of wavy lines of light entering it are resolved as the cornea of a human eye (figure 33). As the features of the face are beginning to be revealed, the eye blinks, and the sequence ends.

Echoing Kees Boeke's desire to impart "a sense of scale" to his readers, this cosmic zoom for an era of mass surveillance makes no bones about the fact that it is a lesson in the trans-scalar logics of the database.[46] We start, innocuously, at a 1:1 scale, Snowden at one end of a mediated chain, a banker who might be funding terrorists at the other. Their relationship

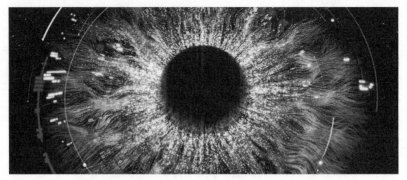

Figure 33 ▸ A planet of data or the window of the soul? Frame from *Snowden* (2016).

is asymmetrical, but their scale is not. Using the logics of social media and the aesthetics of cyberpunk, this sequence demonstrates how quickly an acquisitive algorithm scales up the scope of its dragnet. As this process continues, as we zoom out to ever-greater scales from within the database, the individuals who appear as targets become increasingly arbitrary, with no narrative connection to the original subject. This cosmic zoom, then, while ranging over a banal milieu and employing a recursive structure like its counterpart in *Limitless*, uses the aesthetic of the zoom to visualize the primary bifurcation between the continuous, narrativized self of memory and the scale-unstable organization of what Matthew Fuller calls "flecks of identity," or "the elements of any life that are perspectivally available as 'behaviors' to a particular surveillant scale."[47] Flecks of identity are the limited signatures or components of identity accessible to surveillance systems, which are themselves scale-limited. In the terms of this book, that is, medial systems stabilize certain scales and thereby resolve particular features (subjects/objects) within a narrow band of the scalar spectrum. They also, however, enable scaling, or trans-scalar operations that compose entirely new entities. *Snowden's* cosmic zoom captures exactly this process by which flecks of identity from one scale are aggregated and recomposed into the increasingly expansive mass subject(s) of storage. The result is a scalar expansion as well as an increase in density, a rhizomatic inward growth of connective tissues, like nerve fibers or fiber optics. An ad hoc network of surveilled fragments forms an incomplete, arbitrarily connected totality, what Steve Dietz calls the "dark side" of the database imaginary, "where every shred of information is connected with every other and where total information equals total control," a polarity reversal of the panopticon, from centralization to lateral distribution.[48]

The recursive nature of the *Snowden* cosmic-zoom sequence enables

Figure 34 ▸ A stream of profiles. Frame from *Snowden* (2016).

these two diverging forms of retention, memory and storage, to continually enfold one another. Individual fragments of narrativized life emerge as flecks within data streams functioning at scales unable to resolve the classical subject. The black-hole hemisphere of a darknet database population is scale-collapsed into an eye, both the most human and interiorized of synecdochic scales (the "window to the soul"), and one of the most effective biometric "signatures" of database identity.

One version of the self has an interior narrative, the other is a resource, sitting in a database, ready-to-hand, always a potential target (figure 34). Sandra Robinson has suggested that we conceive of this relationship as that between a subject and her "digital doppelganger," or "a proximal data object that can *do* something in the context of a decision-making apparatus."[49] Ultimately, in the relational database, and in the cosmic zoom, these two forms of identity fold into one another, the narrative ready to be reconstituted from fragments unrelated to their original context, the multiscalar flecks in the database available to reenter the continuous stream of self-narrativization. As Lisa Nakamura has argued:

> Surveillance is a signifying system that *produces* a social body, rather than straightforwardly reflects it. Fingerprint scans, retinal scans, facial recognition, software images, and CCTV images are digitally produced visualizations, virtual spectacles or surfaces that represent a compellingly limited, easily manipulated identity or profile."[50]

In other words, such images, produced by digital surveillance infrastructures, mediate the scale of identity, not only from the perspective of the database, but from that of the surveilled entity. These dissociated identities, torn asunder by scalar difference, reunite in the database as it becomes lived experience, as it becomes productive. "The construction and deployment of databases are part of a political project of identity formation and regulation—they augment without replacing the visual image as the medium of identification."[51]

This is, perhaps, the most disturbing truth revealed/produced by the postdigital cosmic zoom: that memory and storage, diverging most radically in the domain of scale, are *converging* in the domain of medial aesthetics as a new drama of resolution. The weird negative planet of the surveillance network, both black hole and monstrous eye, remembers everything; it couldn't forget if it wanted to. And that memory is my memory; it makes me who I am. Do you remember yourself before the eye? If we stare into the abyss together, what scales will the abyss reflect back into us?

MY DAUGHTER WILL BE A PLANET: THE SCALE-RECURSIVE SELF

In the world of software programming, there is no more vaunted technique than recursion. It simply means code that references itself in the midst of its algorithmic unfolding, generating a mini scalar paradox: its whole is contained within its part. Recursion is an alternative to iteration. Instead of creating some function that operates at the finest grain of your dataset and setting it loose to repeat the process again and again until it finds its result through brute force, a recursive solution sets up a loop, referring back to itself over and over. For example, if we want to ensure that a barrel of apples is red, an iterative approach would go something like this: "Check apple 1 for redness. If not red, paint it red. Check apple 2 for redness. If not red . . . etc." A recursive approach would code a checking routine and a painting routine. Instead of specifying that all actions be performed on every apple, the apples would be sorted according to color and then painted. The protocol would simply refer back to the checking routine and the painting routine without giving a complete set of brute-force instructions. In this way, the procedure would become a loop: first apply X to an apple, then, depending on the result, apply Y. Once a particular result is obtained, go back to X for the next apple, and so on. The procedure becomes a loop, constantly returning to parts of itself. Each time it repeats, however, it is in a slightly different context: a new apple, a new set of freshly painted apples in the barrel, etc. Thus recursion repeats but with a difference: each stage of its operation produces different conditions. It faithfully repeats itself but always somewhere else. James Coplien argues, in a classic piece on the subject, that recursion involves "unrolling time into space," a coding practice in which "we can find beauty."[52] The programmer's disdain for iteration and reverence for recursion is captured in the classic coding slogan, and the title of Coplien's article, "To iterate is human, to recurse, divine." Recursion has the appearance of autonomy: it is created, set in motion initially, but then it sustains itself by making repeated reference to itself (as process). As philosopher Yuk Hui argues, "Recursion presents a form in which the infinite is inscribed in the finite; such an infinite is always an approximation, since in the world of the infinite there is no longer difference in quantity but only in quality."[53] Recursion is always on the razor's edge, then, between the internal and the external, self and other, infinite and finite, qualitative and quantitative. Recursion is a process of systemic aggregation capable of integrating primary and secondary scalar difference.

As this chapter, and recent media history, have progressed, the cosmic zooms found therein have become increasingly recursive. The database logic of scalar transformation through collection and query has molded the cosmic zoom into a form that incorporates both radical scalar jumps and lateral excursions within particular scales in loops that continually re-animate these basic functions, reintroducing already-encountered scales from noncontiguous points on the scalar spectrum. But true to the form of digital (code) recursion, these scales return with a difference: not only has their order been scrambled, but the mode of encounter continually morphs. The third encounter with the brain in the *Limitless* zoom is not the same as the first; it is no longer resolved against the scale of the city and the scale of consumer capitalism, but now against the scale of its basic processing unit, the neuron. The eyeball is banal as a feature of numerous human faces early in *Snowden*'s zoom, but re-resolved against a darknet of global surveillance, it evokes two scalar vectors at once: surveillance media becoming a subjectless perspectival optic, and the human optic nerve becoming a scanning correlate of nonhuman surveillance infrastructure. In all such cases, the database-driven cosmic zoom's recursion reflects mutations in our media culture that have profoundly redistributed agency and altered the functions of trans-scalar access.

Benjamin Bratton has argued that planetary-scale computing necessitates "the *addressing* of every 'thing' therein that might compute or be computed" and that "in many cases, the geography of this addressing bears little or no resemblance to the physical proximity of one addressee to another in physical spaces." The modality of address, then, enables a radical noncontiguity of information transfer, and therefore of action. Moreover, because digital systems suffuse many scales, these noncontiguous operations become scaling operations. As Bratton notes, digital systems enroll entities "into a scope of addressability across and between natural scales, from the infinitesimal to the astronomic, and across natural tempo, from instantaneous to geologic duration," in a form of "deep address" that defies human experience or logic.[54] Such systems do not magically appear and effortlessly link scales. Every scale must be stabilized by media platforms through dynamics of resolution. Nevertheless, the simultaneously decompositional and ontogenetic potential of digital media is clear when we trace the recursive loops it establishes and compels. As access becomes imbricated in and even indistinguishable from address, scaling becomes the most important function not only of media but also of identity.

This is not to claim that scalar recursion eliminates subjectivity. Rather, the recursive form arises from database query as a simultaneously aggregative and compositional force. Scanning flecks of identity and tactically addressing some subset of them from a database, correlated with a particular set of scalar coordinates, produces sites of address that call forth subjectivity, both as a canonically defined variable (address of/as the classic liberal subject, what Louis Althusser refers to as "interpellation") and as a more speculatively arrayed function: address of a collective entity, such as a political group or a cluster of recorded actions that may or may not "belong to" a single human subject.[55] The subject, understood as primarily non-singular, as a function rather than a form, remains the driver of much scalar recursion. A subject navigates a scalar milieu, and thus continually returns at multiple scalar coordinates, driving the cosmic zoom as a medial quest for access that cannot help but to also produce new scales of address. The recursive cosmic zoom, then, is a medial encapsulation of the relationship between subjectivity, database, and scale in contemporary digital culture.

This form of memory enabled by the database represents a marked shift from Kees Boeke's scalar retention of the sociocratic subject. Instead of an accretion of scalar difference in the form of ecological detail, the neoliberal subject is composed of recursive moments of address. Both forms of memory are multiscalar, and both disrupt the standard memory of self-narrative. Their primary difference, however, lies in the relationships they establish between the subject and its milieu. Scalar retention expands the potentials of trans-scalar encounter and internalized alterity, while leaving the singularity of the subject intact. Recursive address flattens ecological detail across scales, while causing the subject to confront itself as its own milieu.

It is perhaps apropos that the first filmic recursive cosmic zoom appeared in one of the first films to feature 3-D CGI: *Futureworld*, directed by Richard T. Heffron (1976). The sequel to Michael Crichton's *Westworld* (1973), the film opens with a dramatic zoom into a human eye, revealing in its interior the protagonist's naked body. The camera continues to zoom into the prone body's face, then into its eye, until it matches the contours

Figure 35 ▶ Recursive zoom in opening sequence. Frame from *Futureworld* (1976).

of the first eye (the cornea of which remains superimposed on the screen). The camera then zooms out to reveal the same body, played by Peter Fonda (figure 35). This recursive structure seems excessive (why split the zoom into two vectors just to portray the same body?) until the subsequent narrative reveals its greatest plot twist—Fonda's brain has been scanned and replicated in an identical robotic body, designed to replace the biological version—and its true theme: artificial life. How do you tell an android from a human? By detecting its recursive programming? Following its loops? The human has loops too, patterns discretizable into database entries.[56] As Wilfried Hou Je Bek notes, "Recursion is behavior made up from the daily routines of life."[57] The human subject is not so much a unified subject as it is a subject function. Database media recursively calls this subject function, addressing it to various scalar coordinates. This is why the recursive cosmic zoom, which visualizes this interplay between subjective perspective and scalar transformation, is born with digital media and thrives within its expanding networks.

Lucy, the film that concludes this chapter, announces from its first two shots that it will explore both spatial and temporal scale: a pulsating blue cell (figure 36) divides in two and an early hominid (which we soon discover is "Lucy," the fabled "first woman" of the human species) drinks at a watering hole. Before long, the modern human Lucy, played by Scarlett Johansson, has been forced into the role of a drug mule by a Taiwanese gangster and has accidentally been exposed to a massive dose of the fictional drug CPH4 (supposedly the synthetic version of a hormone produced by women in minuscule amounts in their sixth week of pregnancy,

Figure 36 ▸ Frame from opening shot of *Lucy* (2016).

which acts like an "atomic bomb" for their fetuses). In a repetition of *Limitless*'s conceit, CPH4 enables the user to access increasingly larger portions of her brain. In contrast to *Limitless*, however, the real subject of *Lucy* is human subjectivity, which is explored at many scales, from the cell (described as handing down knowledge and learning through time) to the species as a whole, and to the cosmos beyond. While the film figures the trans-scalar encounter in chiefly biological terms, its transposition of the concepts of knowledge and transmission are clearly derived from the database form. "There are more connections in the human body than there are stars in the galaxy. We possess a gigantic network of information, to which we have almost no access," lectures Professor Norman (played by Morgan Freeman, almost twenty years after he narrated *Cosmic Voyage*), scale-inverting Paracelsus's notion of the microcosmic body. Norman suggests that information does indeed inhabit a scale, and that in this sense the human brain outscales its milieu in both space and time. In other words, the human brain is a database. This form of the database, however, is not packed with information obtained secondhand—that is, the sort of information gleaned through rote learning or surveillance—but rather information that is synonymous with and indistinguishable from direct experiential contact. In one scene, for example, Lucy touches her roommate and instantly visualizes the interior state, history, and future potentials of her body, concluding with a long prescription for lifestyle change to recover the health her patient doesn't know has been compromised.

Lucy, as database, progressively increases her contact with the world around her, accessing ever-greater scalar alterity in her milieu. As this process continues, however, she finds that her identity changes radically: "It's like all things that make us human are fading away. . . . It's like the less human I feel, all this knowledge about everything, quantum physics, ap-

Figure 37 ▸ Becoming database. Frame from *Lucy* (2016).

plied mathematics, the infinite capacity of the cell's nucleus . . . they're all exploding inside my brain, all of this knowledge." Though she is being pursued by the gangster, who wants his CPH4 back, and facing imminent death, she describes herself as feeling no pain or fear. Though she wishes to pass her newfound knowledge on to the human race through Professor Norman, she is quite willing to let go of her own human identity. "Now . . . I see things clearly and realize that what makes us *us* . . . it's primitive. They're all obstacles."[58] Lucy, like her primitive namesake, is the first member of a new race capable of transcending its biological provenance: Nietzsche's *Übermensch* for a postdigital age. Indeed, in the final sequence of the film, she literally becomes a database containing knowledge of the entire universe (figure 37), conveniently packaged as a USB flash drive.

Lucy's final encounter with the scales of reality, in which she completely and irreversibly overcomes her scale-limited human nature, takes the form of a cosmic zoom. In this case it is composed of three cosmic sequences, intercut with a final, surreal encounter between the Taiwanese gangster and Lucy at the Sorbonne—a sequence that provides an ironic counterpart to Lucy's transcendence, a recurring reminder of the barbaric limitations of human subjectivity concerned "more with having than being."[59] In the first sequence, Lucy gains the ability to transport herself in time and space, appearing at a series of locations, ending at Times Square in New York City. Here she stops time, and, with a swipe of her hand, sends it flowing backward. Times Square travels back in time through the twentieth, nineteenth, then eighteenth century, until Lucy encounters several Native Americans on horseback, who regard her with some surprise. Traveling further back, she finds herself in what appears to be the Jurassic age. When a violent dinosaur attempts to attack her, she shifts both time and space to appear before Lucy, the "first woman" we encountered at the

Figure 38 ▸ Cosmic spermatozoa. Frame from *Lucy* (2016).

beginning of the film. They regard each other intently for a moment, then touch their index fingers together in a gesture mimicking Michelangelo's fresco *The Creation of Adam*. The moment of contact initiates a sudden vertical zoom, in classic cosmic-zoom fashion, into orbit around the earth. In this second sequence, Lucy herself doesn't appear, but time begins to run backward again, allowing us to witness the formation of the earth's continents in reverse, then its early, fiery beginnings as asteroids and planetoids collide to form its mass. In the third segment of the cosmic zoom, as Lucy is approached from behind by the gangster, the camera zooms through exotic, unfamiliar nebulae. Things get much weirder, however, when a rain of forms resembling jellyfish emerge from a stellar cloud (figure 38). Zoomed out further, they are resolved as spermlike forms, numbering in the thousands, descending upon and entering into something like an interdimensional wormhole.[60] As the camera travels through this wormhole, a number of blue cells (of the sort that appeared at the beginning of the film) begin to merge. When the wormhole finally bifurcates and meets itself, it is at the exact instant the final cells merge into one, and Lucy's form as Scarlett Johansson completely vanishes. Physically, she presents herself as a flash drive, containing the sum of her knowledge, passed on to Professor Norman. Subjectively, she diffuses into the world around her, announcing to her friend, the detective, "I am everywhere."

The cosmic zoom, then, is the final stage of Lucy's identity transformation from a human subject to a trans-scalar one. Not just constructed from a database, she has become the database itself. And in so doing, she has forced the database also to become something else. It is storage, yes, but also a direct manifestation of scalar difference, of the trans-scalar encounter and all entities involved in that encounter: primitive Lucy, modern Lucy, the Native Americans, the dinosaur, the intergalactic spermatozoa,

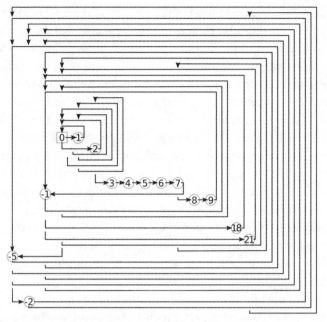

Figure 39 ▸ Recursive diagram of *Lucy*'s cosmic-zoom sequence. Scales are labeled as powers of 10 meters. The sequence begins with 10^0, or one meter (indicated by the square).

the earth, and the computational infrastructure that she subsumes into herself in the Sorbonne laboratory that serves as the site of her embarkation. That is, Lucy harnesses the trans-scalar, ontogenetic properties of the relational database that I have been tracing in this chapter and pushes them beyond both representation and their technologically limited forms of address. The ultimate question the film asks is this: When trans-scalar access is increased, what do we do with all of the resulting trans-scalar knowledge that no longer fits into disciplinary categories? This knowledge, I suggest, becomes recursive: no longer a representation of the world observed, it becomes a participatory resolving cut that cannot but form a new subject (figure 39).

Like figure 31, figure 39 is a diagram of scale recursion. *Lucy*'s entire cosmic-zoom climax can be traced by starting at the square initial scale (the scale of the human body) and following the lines top to bottom, and left or right in the direction of the arrows. Every line below the initial scale represents a progression to a different scale, while every line above the initial scale represents a return to a previous scale—many jumps involve both. Unlike *Limitless*, however, what is explored here is less a multiscalar milieu than a multiscalar subjectivity. Like the cognitive wiring of a subject distributed in some nonmolar medium such as silicon or rhizomatic root structures, *Lucy* expands along both scalar vectors at once. She takes on the role of a planet while also birthing microscopic cells that dance with her/their liveliness. Is this the diffusion of identity, the expelling of a life force outward in all directions, to be dissipated into the uniformity of cosmic background radiation? The scale-recursive diagram indicates otherwise: as wildly scale-divergent as these circuits are, they return again and again to a subjective center that holds the entire set of circuitry in metastable tension. Scale recursion is a continual return, but always with a difference. Each return simultaneously reactivates the scalar affordances of a previous node (including all of its trans-scalar links from previous activations) and participates in a further progression. Rather than simply propagating through a trans-scalar ecology, then, scale-recursive processes stretch to the material substrates of alterior scales at the same time that they cycle through their familiar or home scales, producing variation as they loop. As a diagram of an experimental form of subjectivity, then, figure 39 is a schematic of multiscalar becoming.

The engine behind *Lucy*'s scale-recursive subjectivity is the database form. Instead of encountering the monstrosity of the database as merely a form of control in our digital milieu, a threat to the autonomy of the human individual as figured in *Snowden*, *Lucy* suggests that a greater engagement with the world might, as a vector, move *through* the database rather than away from it. Thus, while the cosmic zoom can and should be deployed effectively to visualize the horrors of digital surveillance and profiling, it can also point to a path beyond. This isn't too far from a potential Sherry Turkle highlighted in 1995, in the early days of both the Web and window-based graphical user interfaces: "Windows have become a powerful metaphor for thinking about the self as a multiple, distributed system." Instead of playing multiple roles in *series*, which we are forced to do in everyday life, "the life practice of windows is that of a decentered self that exists in many worlds and plays many roles at the same time."[61] While Turkle's hopes that the Internet would usher in a widespread change in human subjectivity thus far remain unrealized—indeed, what we have mostly witnessed in online culture is a retrenchment and amplification of classical human identity, which if anything has become even more hyperbolically defensive of the sanctity of individuated (by neoliberal capital) identity—the necessary infrastructure that potentiates Turkle's shift has in fact assembled itself. The irony is that it has taken the most oppressive form possible: what Mark Poster describes as the "superpanopticon," a database-driven form of inscription that composes an entirely new form of subjectivity, "producing individuals with dispersed identities, identities of which the individuals might not even be aware."[62] With the affordances of scale theory, however, we can differentiate between primary scalar difference (ontological scale) and secondary scalar difference (mediated scale), and recognize how the inextricable interplay between the two acts as the substrate for both milieu- and identity-formation. This enables us to see, even from within the heart of our database-driven media ecology, the compositional potentials of the database form, and helps to guide media toward a generative (non-scale-collapsed) form of scalar mediation.

My argument in the second half of this chapter has been that the same digital forms that entrap us in our current milieu also deliver the tools to extend that milieu, and thereby subjectivity. The database form of media, as *pharmakon*, contains both potentials because it has fundamentally shifted from access to address. Just as this has opened up new forms of the cosmic zoom, it has opened up new forms of trans-scalar encounter. That is, the move from representation to participation across scales is enabled by the recursive form ushered in by digital code. Mark Hansen, in applying the philosophy of Alfred Whitehead to the ontological mutations of digital media, notes that digital media ontologically expand the world of potential experience:

> *At one and the same time*, twenty-first-century media *broker human access to* a domain of sensibility that has remained largely invisible (though certainly not inoperative) until now, *and*, it *adds to* this domain of sensibility since every individual act of access is itself a new datum of sensation that will expand the world incrementally but in a way that intensifies worldly sensibility.[63]

The structure of this incremental expansion of the sensible is recursive because it proceeds by addressing new scalar sites that thereby become incorporated in digital media's general affordances of address. The more it creates in its process of encounter, the more there is to encounter. At the same time, because database-driven media increasingly address scales and sites other than the human, yet remain themselves (largely) accessible to the human, they become conduits to trans-scalar encounter. But as the self participates in such encounters, it is addressed from and as new scalar compositions, and thus gains the capacity not only to access other perspectives but to occupy (as addressee) new scales as a subject.

That the radically transformative potentials of the trans-scalar encounter are not inevitable I take as manifest. Resurgent human individualism, driven by such diverse forces as neoliberal capital, neofundamentalisms that oppose neoliberal capital, conservative forces that seek to enforce the purity of dominant forms of subjectivity, and liberal forces that seek to protect and encapsulate the increasingly fragile humanist ego—everywhere confronts us with the twin forces of human(ist) exceptionalism and identity essentialism. Somewhere in the middle of the fray is the recursive cosmic zoom. Its continued drama of resolution contains all of the players and all of the scales in play. You can't play without getting played, you can't access without being addressed, you can't encounter other scales without encountering yourself. Nevertheless, the recursive self remains unrealized. What would it look like?

Recursion is a scaling strategy, not a continual return to one scale but, rather, the repetition of an address function that continually traverses new scales. Accordingly, the recursive self would be multiscalar in its perspectival register. Like the train-hopping itinerants of American folklore, the recursive self would be, by necessity, scale-nomadic. But also like them, it would be free to follow the trans-scalar infrastructures of its milieu, to hop a ride when necessary and thereby trace a continuity between noncontiguous scales. While classical humanist subjectivity approaches scalar difference from a fixed perspective, resulting in the forms of disciplined access and consequent scalar collapse traced in chapters 4 and 5, the recursive self would be capable of maintaining scalar difference in productive tension (between self and other as well as between self-for-self and self-for-other), thus preventing the foreclosure of the trans-scalar encounter. This would come at a severe cost, of course: the endangerment and sometimes complete abandonment of the comfort of narrative continuity (memory). The sort of experiential continuity that obtains across scales is, as *Lucy*'s cosmic zoom demonstrates, dissociative. This dissociation, medially traced by database-driven cosmic zooms, is fundamental to recursive database queries, which produce new assemblages that do not map onto the scalar coordinates of classically defined objects. The result, from the perspective of the querier, is that one keeps encountering oneself, but not as narrative continuity; rather, as a part identity (certain flecks of identity aggregated with those from other sites) or as encapsulated within a larger assemblage (produced by the query) or, often, both simultaneously. I discover that I am part of an assemblage that has only just come into being, defined by consumer patterns or political interests or governmental paranoia. Or else I discover that that I am a geolocative trail, a series of debits in an account, a collection of tracked gestures. A form of identity that embraced and found a way to persist through such dissociation and transformation could also incorporate these radically ontogenetic affordances.

Just as the database form both captures and composes, recursion as a modality both repeats itself and discovers ever-new worlds. As Yuk Hui

Figure 40 ▸ Planetary ovum as simultaneous subject, milieu, and process. Frame from *Lucy* (2016).

emphasizes, this is effectuated through a process of incorporating more information into a system, which in turn changes not only the system's relationship with its environment but its own character:

> Information provides a conceptual tool applicable to all orders of magnitude, whether it be the micro-level of electrons or the macro-level of steam engines. A contingent event brings information into the system and produces a signification to certain elements in the system or the system as a whole, which in turn triggers a new process of individuation.[64]

The recursive self would maintain its integrity as an individual assemblage over time through this loop structure, a continual calling of its previous functions. These would always operate in new conditions, however, thereby dashing any fantasy of prediction and control (contra *Lucy*). Wilfried Hou Je Bek cautiously describes recursion's perpetual-motion machine as an ideal loop: "The LOOP is the powerhouse of worlds imagined in silico: the sweat-free producer of matter and time."[65] The recursive self I'm outlining here, as a particular loop through medial infrastructures and across scales, differs from the ideal version imagined by software engineers in that it is far from sweat-free. It obeys the laws of conservation of mass and energy, but it does indeed produce (or rather, recompose) matter and time. Far from the neoliberal dream of unfettered, resource-free production, this is a messy, sticky process; messy because it admits of no central control, and sticky because everything that is created incorporates the creator(s) and the site of creation (figure 40). You may give birth to a planet, or you may be the planet, or you might just get caught in a downpour of cosmic, interdimensional sperm. *The Hitchhiker's Guide to the Galaxy*'s sage advice is apropos here, as elsewhere: "A towel . . . is about the most massively useful thing an interstellar hitchhiker can have."[66]

This book opened with Alice falling down a rabbit hole and discovering the scalar delights and horrors of the Anthropocene: first awareness, then blundering encounter, and finally disciplined access. What she couldn't do, despite her trans-scalar encounter, was dissociate from her own subjective scales. She remained human, all too human, despite encountering scalar others. She could not become recursive. Lucy, as planet, as cell, as database, completes Alice's journey through Wonderland. This is not to say that we can all become planets and thereby "solve" climate change or the surveillance state. Rather, Lucy is an invitation to loosen our grip on the human subject, not merely through consciousness of radically nonhuman, multiscalar ecology, but additionally through an openness to reconceiving subjectivity as existing *between* scales, an openness to dissociation, to inhabiting the loop in which scalar processes compose the subjects that compose scales that compose new subjects . . . Wonderland not as environment but as recursion between primary scalar difference and secondary scalar difference.

In classical Greek drama, *anagnorisis* marked the protagonist's acquisition of knowledge critical for his or her full self-understanding. To put it mildly, as fans of *Oedipus Rex* know all too well, *anagnorisis* was none too flattering. The recursive self, tracing the circuits of database-driven media, is doomed like Sisyphus to continually return to the same encounter. In this case, however, the encounter is with oneself and is never an exact repetition: it marks not a return to privately staked-out territory (or the arrival of bad news via messenger), but rather a scalar jump to a new and unfamiliar site of address. In this context, *anagnorisis* takes on a slightly altered function. Via the database form, one learns something about the construction of the self. Beyond the social media "likes" and "friend" networks, one is apt to discover, from database-driven media, both that one is a product of neoliberal norms and economic systems within which all choice is revealed as illusory, and that one is composed of flecks with their own autonomous histories and provenances. I am the product of storage as much as I am of memory. I am a site of global distillation as well as the expanding assemblage of molecular actions and accretions.

Unlike for Oedipus, scale-recursive *anagnorisis* comes, at least partly, from within. No messenger arrives, or even if she does, you accompany her, you see through her eyes. This knowledge is not an imposition on the integrity of your tightly woven kingdom but rather *news from another scale*, and however bad it may sound, you welcome it because, for the recursive self, navigating scalar difference is not something a self *does* but an opening to becoming milieu, while the milieu becomes you: a cosmic scalar spiral.

coda

dwelling in the scalar spectrum

Phylis Morrison published a short review of Kees Boeke's *Cosmic View* in 1979, two years after the release of *Powers of Ten*, in which she highly recommended the book, for children in particular, because it "places the reader at home in the worldview of our epoch."[1] Where is this home in which Morrison suggests the reader is placed? It is located, apparently, within a worldview. But can there be a single worldview for our epoch? The official worldview of the Anthropocene? The digital age? Late capitalism? Morrison's supposedly totalizing worldview is, I think, rather literally a view of the world: one that encompasses all scales, from the quark to the outer bounds of the universe. Of course, such a view can only be assembled from fragments.

In chapter 2, I argued that *Cosmic View* affords not an overview of the entire universe but a discontinuous series of resolving cuts, which Boeke takes great pains to present as medial slices that connect the reader to a virtual surface by negotiating a series of relationships between that surface and the physical pages of the book in her hands. Thus Boeke's attempt to impart "a sense of scale" amounts, not to a totalizing overview, but to a self-reflexive mediating machine that emphasizes the ineluctable tradeoff between field of view and detectable difference in a drama of resolution. The result is that the reader gains insight into the processes of scalar mediation and the stabilization of discrete scales, but gets no closer to a universal perspective. Instead, the titular cosmic view takes the form of a virtual horizon, enabled only through a kind of medial memory recorded on the retina and the tips of page-turning fingers. It is precisely because a universal overview is impossible that such scalar memory is necessary. Boeke's book serves as a primer in scale literacy, promising both the metaknowledge of scalar mediation and the possibility of scalar openness, of receptivity to scalar alterity and humility in the face of that which each resolving cut excludes.

This could well have been the legacy of Boeke's work were it not for the turn that the cosmic zoom took in later decades, a turn that would vastly increase its cultural influence and fundamentally alter the way we think about scalar alterity. I have documented the long and arduous processes behind this medial project as it progressed, from the 1960s to the 1990s, toward the production of a smooth scalar continuum that could deliver the illusion of witnessing the unfolding of all scales as contiguity in space. The tradeoffs of resolution and the transparency of the medial processes of scale stabilization became ever more occluded as, conversely, the thrill of scalar access became all the more palpable. When Phylis Morrison retroactively reviews Boeke's pioneering work, then, it is through the lens of her own involvement in translating that work into a vehicle of scalar collapse, which in turn helped to catalyze a movement of technoscientific scalar access that continues unabated to this day.

As Heidegger reminds us, this is nothing new. "The fundamental event of the modern age is the conquest of the world as picture."[2] As I discussed in the last chapter, this is not a *particular* picture but the metaphysical formulation of the world as pictorially capturable, as an object marked by continuity and thus representable totality. The triumph of technoscience, its wondrous smashing of barriers to access, coupled with its occlusion of its own processes of mediation, makes possible the naturalization of scale into a linear extension of the human milieu: technoscientific manifest destiny and the "Wild West" as frontier territory rolled into one. Pan-scalar humanism as an ideology of scalar access reinvigorates the liberal human subject as privileged master of its domain, now in the neoliberal context of total market enrollment. It is not coincidental that as European colonialism faltered after World War II, a hylomorphic, trans-scalar colonialism began to take its place. It is a question of resources: when they have been fully exploited on one plane, new planes—territories—must be opened up for exploitation. Scalar prospectors made the discoveries, but to get at the goods, capitalists soon learned that they must dig mines not only deep into the earth but also along other vectors, deep into the scalar spectrum. Morrison's suggestion that Boeke's book—and by extension the entire cosmic-zoom project—"places the reader at home in the worldview of our epoch" is thus a suggestion that the cosmic zoom puts us at ease dwelling within a world picture, a representable totality.

This book has examined the cosmic zoom as a privileged medial form that takes scale itself as its subject and is thus particularly revealing of scalar politics, but it is by no means the only form of scalar mediation. Indeed, all acts of mediation are resolving cuts that enable trans-scalar encounter through the negotiated alignment of disparate surfaces. This book has therefore been an exploration of a key function of media that is usually overlooked or is only hesitantly and partially approached by media scholars and popular accounts. I hope to have more systematically, yet also more suggestively, opened a dialogue about the importance of scale in all acts of mediation, and the profound consequences of seemingly innocuous medial flirtations with scalar alterity.

Analog and digital media offer different scalar affordances and pitfalls. As articulated in this volume, an analog universe is one of temporal continuity transmuted into a spatial contiguity that collapses scalar difference in service of the liberal humanist subject (which acts as the gravitational center around which this contiguous trans-scalar space bends). Ironically, analog media cannot help but to show their seams between scales, however tightly interlocked their arcs of resolution. Digital media are typically far more effective at hiding their own seams, and thus with the advent of the pixel, the analog universe becomes far more effective as an embodied concept—that is, as an ideologically driven medial argument. Nonetheless, the digital cosmic zoom has been far from a smoothed, perfected analog zoom. As I have argued, the affordances of the database have mutated the digital cosmic zoom, liberating it from the slavish rollout of an endless, contiguous space. Infinite zooms through precise scalar territories and recursive scaling have reintroduced difference into medial models of scale, hinting at the seething powers of scalar differentiation inaugurated by digital media.

In practice, scalar mediation is very often a mediation *between* the analog and the digital. That is, sites of scalar transformation often harness differentials between analog and digital entities to effect their jumps; alternatively, digitization as a process is inherently trans-scalar: it scans and encodes certain target scales that frequently function below the scales of the self-narrated individual, while reclassifying and aggregating such flecks into larger virtual assemblages. This, I have argued, is the primary scalar function of the database form of mediation, which follows the logic not of representation, but of composition. At the heart of the compositional mode of database-driven scalar mediation is the query, an interfacial encounter with the networked entries of any database, with the potential to resignify previously unrelated forms and scales into new ones.

Though it has been beyond the scope of this book, a scalar analytic could shed new light on mechanisms of social and economic injustice usually approached from liberal humanist frameworks: racism, sexism, classism, nationalism, and speciesism all incorporate and rely upon scalar constellation. They are productive processes that organize forces, composing scales through which populations and identities can be managed and contained. Power is notoriously difficult to track and analyze precisely because of these scalar dynamics. Tactically it produces magic shows that enable force to materialize and then quickly shift to other scales, rendering it invisible to its victims. Perhaps even more frustrating, there is no easily identifiable perpetrator behind these tactics: power arises out of scalar difference itself as an autopoietic diagrammatics of control. The individuals and groups who occupy privileged scalar nodes within the dynamic networks of capital and knowledge benefit the most from these asymmetries, and almost invariably labor to perpetuate the structures that placed them there, but they cannot be considered architects of power. Power is organized at many scales, and its orchestration cannot be reduced to a single one; the myth that it can has led justice seekers on a wild goose chase. It is my hope that the analysis of scale developed in this book may open up new potentials for the tracking of power across and through its scalar articulations, and thus new forms of resistance commensurate with its most pernicious manifestations at all scales. In the contemporary milieu, the more virulent strains and simmering undercurrents of trans-scalar, neoliberal capital are beginning to resemble neofacism. The full scalar implications of neoliberal capital and far-right populism, however, are still in need of elaboration, for our old theories see only limited slices of reality, doubly hampered by their scalar biases and their inability to resolve scale as a bias.[3]

Kees Boeke's creative response to fascism, which he preferred to call "narrow provincialism," opened a path toward a radical politics of scale, even if it was quickly colonized by the proto-neoliberal forces of techno-scientific boosterism and (human-centered) design. Ironically, the twentieth century's most conservative, anthropocentric model of scale was inspired by a radical, anticapital anarchist. Long before digital media's affordances of the pixel, network, and database would both increase the political stakes of scalar mediation (as astonishingly effective tools of capital) and chart a new path forward for the liberation of scalar difference beyond provincial humanism, Boeke envisioned a different legacy of the cosmic zoom. In December 1957, directly after the initial publication of *Cosmic View*, he began work on another book, a scalar sequel entitled *The Great Race*.

Boeke intended *The Great Race* to depict scale in relation to speed. Repeating the scalar-jump form of *Cosmic View*, it would progress through ten orders of magnitude, recording faster and faster velocities as they approached the speed of light, the definitive end of this scalar spectrum. Where *Cosmic View* made forty-two jumps, *The Great Race* would make only ten, but each scale is populated by a significant range of denizens.

Velocity is a provocative scalar metric because it incorporates both spatial and temporal axes, and like some of the digital projects discussed in chapter 6, it enables the formation of scalar types (generic forms) that cut across spatial contiguity. But while database-driven media can create new scalar aggregates from nonsimilar objects that occupy the same physical region of the scalar spectrum, the lateral aggregations of *The Great Race* would have scrambled size as well. The aircraft carrier *Lexington* is found on the same page as a jackrabbit.[4] As velocity scale increases, the size of objects does not necessarily follow suit: the fastest object in *The Great Race*, found on the final page, is the electron. This scrambling of size through aggregative functions articulated to alternative metrics not only looks forward to database-driven scalar media but also serves to highlight what I have called the generic scalar event, a typical form of action that contributes to stabilizing a scale as a milieu. Velocity, or speed, is in this sense a generic quality or capacity: it defines both a capacity of movement and a set of ecological protocols of interaction. Objects traveling at differential speeds meet all of the time, but only momentarily. Those meetings can be generic scalar events, as when electrons accelerated by a cathode ray tube habitually collided with a human-scale television screen, serving to define a new scale and scalar function in the mid-twentieth century: the living room. But creative medial aggregation can also reveal the strange or speculative affinities of objects that habitually travel at similar velocities, producing new virtual scalar milieus.

Boeke's mockup of *The Great Race* includes ten scalar-velocity domains on ten two-page spreads, progressing from 0.01 miles per hour (the minute hand of Big Ben) to 186,000 miles per second (the speed of light). One of the most striking characteristics of the project is the diversity of the entities collected in its pages. Humans, and even human technology, account for only a small portion of these scalar denizens. Instead we encounter blood in an artery, a coral snake, a stratospheric hurricane from 1936, the planet Neptune, and colliding galaxies (hearkening forward to *Cosmic Voyage*). What, we may be tempted to ask, are the shared characteristics of these radically disparate objects? What are their potentials for encounter? Many scales are subdivided via thick rows into aerial, terrestrial, and aquatic domains, further enhancing the evocation of scalar milieu, or scalar coordinates that serve virtually (in media form) to aggregate and (in elemental form) to effect encounter at an ecological site.

Boeke and illustrator Els de Bouter worked for three years on the project. Boeke first described it to Richard Walsh, his editor for *Cosmic View*, in December 1957, and Walsh was eager to publish it.[5] In 1960, however, Boeke wrote to Walsh that "I must give up the proposed publication of 'The Great Race'" due to Els's failing health (she had told Boeke she could no longer travel to Abconde, where he lived, to collaborate on the book) and, a new development, "my loss of memory."[6] Boeke had not previously mentioned his loss of memory in his correspondences with Walsh, and does not elaborate further.

Boeke had previously met two visiting US teachers and shared details of *The Great Race* with them. In February 1961 one of them, Hyman Kavett, wrote to Boeke and proposed that he finish the book Boeke had abandoned.[7] Boeke agreed, and over the next year, sent his research, the latest mockups, and all of Els's drawings to Kavett. Unfortunately, Kavett never finished the project. Boeke died in 1966, and *The Great Race* was all but erased from history. It is not mentioned in any scholarly or popular works on *Cosmic View* and was unknown even to Boeke's school, the Werkplaats, despite a resurgence of interest in the institution's founders by its current director, Jos Heuer.[8] The drawings and final materials have been lost; only early notes and Boeke's letters on the subject remain with his papers in Amsterdam. Instead of this radically noncontiguous work on scale, posterity would be left only with *Cosmic View*, which, as the pages of this book have documented, was bled of its radical potentials and resignified in media history by *Powers of Ten* and its legacy of pan-scalar humanism.

In Boeke's regretful letter to Walsh, he names his loss of memory as a primary reason for abandoning *The Great Race*. Just five days earlier, however, he had meticulously compiled a nine-page document listing *every correspondence he had ever had with Walsh*, including the date, medium (telephone or letter), and subject matter of the communication. What possible purpose could this document serve? Boeke certainly didn't need this record for the purpose at hand (informing Walsh that he couldn't continue the project). Rather, it seems to have been an exercise in protocological aggregation, the collecting of metadata as both a metonym for the contours of a long-term collaboration and as the composition of a new entity at a new scale (spanning seven years and two books, but reduced to only two individuals), out of disparate traces of previous trans-scalar encounters. Boeke, apparently losing the thin thread of narrative memory that largely defines the continuous human subject, adopts the database form of memory as storage, not as a functional necessity but as an experimental tactic: he produces a new query by way of terminating another, an adequate informatic approach to the trans-scalar milieu in which he finds himself.

Just as Boeke becomes discretized and recomposed at new scales, so too may the rest of us/them/you have these trans-scalar energies liberated from the molarizing forces that seek (with general success) to contain them in the vessel of the liberal human subject. These forces are themselves trans-scalar, but much of their power harnessed as control stems from their ability to reinforce this scalar containment. Instead, I propose that we embark on a journey that begins by following those forces of composition.

We can, I have argued in these pages, mediate scale more deliberately and less rigidly. We can harness scalar mediation as a creative intervention into the ossified infrastructure of our psyches, cities, local ecosystems, nations, and bodies. At the same time, we must remember that scalar mediation is a circuit that contains two halves: the subjective stabilization and exploitation of scales as knowledge domains and the undisciplined irruption of difference that emerges from every resolving cut, an ontological force that variously enslaves, subjugates, composes, and liberates, even as it remains the object of ongoing negotiation (resolution). The goal of such negotiations, from the posthuman perspective, can no longer be mere access to and assimilation of other scales, but rather a creative and active mediation of our own regimes of resolution and composition. The potential of every trans-scalar encounter can be invested in the liberation of difference and resolution of ecological detail. It is only in this humble, peripatetic way, open once again to wonder and awe, that we can begin to change our species, to build a new home somewhere on the shifting sands of the scalar spectrum.

acknowledgments

The scale of this project has made for a journey sometimes ecstatic and sometimes treacherous. I am deeply grateful for those who have chosen to travel the scalar spectrum with me, as guides or companions. I cannot enumerate them all, but particular thanks are due to Alan Liu, Rita Raley, Bishnupriya Ghosh, Colin Milburn, Bhaskar Sarkar, Stephanie LeMenager, Maurizia Boscagli, Janis Caldwell, Jeremy Douglass, Christopher Newfield, Teresa Shewry, Enda Duffy, Janet Walker, Patrick McCray, Lindsay Thomas, Susan Derwin, Hsuan Hsu, Michael Travel Clarke, David Wittenberg, Jennifer Waldron, Troy Boone, Annette Vee, Neepa Majumdar, and John Durham Peters. I am particularly indebted to Jessica Horton and Alexandra Magearu for their companionship, support, and impeccable editing and feedback. My warm thanks also go out to Don Bialostosky, Gayle Rogers, and the Department of English at the University of Pittsburgh for their tremendous support of this work. I am also grateful for my anonymous University of Chicago Press readers for their generous, and generative, feedback. Everyone listed above shaped my thinking about

227

scale and elevated this book. A number of institutions also proved indispensable to this research. I'm especially grateful to Jos Heuer, director of the Werkplaats Children's Community in Bilthoven, as well as the Library of Congress in Washington, DC, and the International Institute of Social History in Amsterdam. This book wouldn't exist without the expert guidance of Kyle Wagner, Joel Score, and Dylan Montanari at the University of Chicago Press, for which I am deeply grateful. Finally, I thank my family for their loving support, and for introducing me to this great big world.

notes

CHAPTER 1

1 ▸ Lewis Carroll, *The Annotated Alice: The Definitive Edition*, ed. Martin Gardner (New York: Norton, 1999), 18.

2 ▸ Carroll, *Annotated Alice*, 53.

3 ▸ Throughout this book I use the term *scalar spectrum* to designate the range of possible scales, from the smallest to the largest, understood as a discontinuous, heterogeneous set rather than as the space or time traversed by an abstract scaling operation that would smoothly integrate all possible scales. "Scalar spectrum" can be usefully counterposed to, for instance, "scalar continuum," which designates an interscalar vision of smooth continuity and infinite scaling, characteristics often imparted, as we shall see, by particular cosmic zooms.

4 ▸ Michel Foucault, *The Order of Things: An Archaeology of the Human Sciences* (New York: Vintage, 1973), xxi.

5 ▸ Rob Nixon, *Slow Violence and the Environmentalism of the Poor* (Cambridge, MA: Harvard University Press, 2011), 3.

6 ▸ Ant-Man, a Marvel superhero who has the ability to change almost instantaneously from standard human to ant scale, debuted in 1978 and never had his own comics series. The film *Ant-Man*, directed by Peyton Reed and released by Marvel Studios in 2015, nonetheless proved extremely popular. Ant-Man has since been featured in two more

blockbusters, *Captain America: Civil War* (dir. Anthony Russo and Joe Russo, 2018) and *Ant-Man and the Wasp* (dir. Reed, 2018). Other recent films that feature trans-scalar encounters in prominent sequences are discussed in chapter 6.

7 ▸ Derek Woods, "Scale Critique for the Anthropocene," *Minnesota Review*, no. 83 (January 2014), 133, https://doi.org/10.1215/00265667-2782327.

8 ▸ Stacy Alaimo, *Bodily Natures: Science, Environment, and the Material Self* (Bloomington: Indiana University Press, 2010), 2.

9 ▸ David Harvey, *The New Imperialism* (Oxford: Oxford University Press, 2005), 43.

10 ▸ K. Eric Drexler, *Radical Abundance: How a Revolution in Nanotechnology Will Change Civilization* (New York: BBS PublicAffairs, 2013), 11.

11 ▸ Marcus Tullius Cicero, *The Dream of Scipio (De Re Publica VI 9–29)* (Schwartz, Kirwin & Fauss, 1915), 29.

12 ▸ Cicero, *Dream of Scipio*, 33.

13 ▸ Cicero, *Dream of Scipio*, 31, 33.

14 ▸ For my initial discussion of scalar collapse, see Horton, "Collapsing Scale: Nanotechnology and Geoengineering as Speculative Media," in *Shaping Emerging Technologies: Governance, Innovation, Discourse*, ed. Kornelia Konrad et al., Studies of New and Emerging Technologies 4 (Berlin: IOS Press, 2013), 203–18.

15 ▸ Siegfried Zielinski and Timothy Druckrey, *Deep Time of the Media: Toward an Archaeology of Hearing and Seeing by Technical Means*, trans. Gloria Custance (Cambridge, MA: MIT Press, 2008), 11.

16 ▸ I was particularly struck by the situation of interlocutors from different disciplines deploying disparate notions of scale when, in June 2015, the European branch of the Society for Literature, Science, and the Arts convened the first large conference in the humanities devoted entirely to the theme of scale. Presenters approached the theme from many different modalities and fields, often reconceptualizing their previous work in terms of scale, but few defined what they meant by scale. It soon became clear that numerous conceptions of scale were being deployed without their provenance being either acknowledged or problematized.

17 ▸ William T. Pryor, "Selection of Maps for Engineering and Associated Work," in *Map Uses, Scales, and Accuracies for Engineering and Associated Purposes: A Report of the ASCE Surveying and Mapping Division, Committee on Cartographic Surveying* (New York: American Society of Civil Engineers, 1983), 28.

18 ▸ "Contour interval and map scale are the principal parameters considered in selecting and/or compiling topographic maps; whereas, scale alone generally governs in the selection from available sources or compilation of the other types of maps." William T. Pryor, "Maps for Engineering and Associated Work by Stages," in *Map Uses, Scales, and Accuracies for Engineering and Associated Purposes: A Report of the ASCE Surveying and Mapping Division, Committee on Cartographic Surveying* (New York: American Society of Civil Engineers, 1983), 11. Contour interval is elevation resolution: the degree of elevation difference that registers as a further contour "ring" on a particular map projection.

19 ▸ Pryor, "Maps for Engineering and Associated Work," 9.

20 ▸ Timothy Clark humorously reminds us of the importance of scale vis-à-vis resolution with the following anecdote: "You are lost in a small town, late for a vital appointment somewhere in its streets. You stop a friendly-looking stranger and ask the way. Generously, he offers to give you a small map which he happens to have in his briefcase. The whole

town is there, he says. You thank him and walk on, opening the map to pinpoint a route. It turns out to be a map of the whole earth. The wrong scale." Timothy Clark, "Derangements of Scale," in *Telemorphosis: Theory in the Era of Climate Change*, ed. Tom Cohen, vol. 1 (Ann Arbor, MI: Open Humanities Press, 2012), 148.

21 ▸ Karen Barad, *Meeting the Universe Halfway: Quantum Physics and the Entangle-ment of Matter and Meaning* (Durham, NC: Duke University Press, 2007), chap. 3.

22 ▸ This is not merely a question of John Locke's famous distinction between primary qualities (those that inhere in the thing itself) and secondary qualities (those produced through sensory organs perceiving the thing). See John Locke, *An Essay Concerning Human Understanding*, trans. P. H. Nidditch (Oxford: Clarendon Press, 1975), 135. Gold nanoparticles suspended in solution produce these varied colors even when perceived at the macro scale. This effect is evident in much of the stained glass of medieval European cathedrals and churches, leading Ratner and Ratner to suggest that "in some senses, the first nanotechnologists were actually glass workers in medieval forges." See Mark A. Ratner and Daniel Ratner, *Nanotechnology: A Gentle Introduction to the Next Big Idea* (Upper Saddle River, NJ: Prentice Hall, 2003), 13.

23 ▸ I am grateful to Jeremy Douglass for suggesting the figure of the layer cake.

24 ▸ Max Born and Albert Einstein, *The Born-Einstein Letters: Friendship, Politics and Physics in Uncertain Times* (Houndmills, UK: Palgrave Macmillan, 2005), 91; David Bohm, *Wholeness and the Implicate Order* (London: Routledge, 2002), xviii; Stephen Hawking, *A Brief History of Time*, 10th anniversary ed. (New York: Bantam, 1998), 14.

25 ▸ D'Arcy Wentworth Thompson, *On Growth and Form*, ed. John Tyler Bonner, abridged ed. (Cambridge: Cambridge University Press, 1992), 36.

26 ▸ John Tyler Bonner, *Why Size Matters: From Bacteria to Blue Whales* (Princeton, NJ: Princeton University Press, 2011), 148.

27 ▸ Thompson, *On Growth and Form*, 48.

28 ▸ Gilbert Simondon, *On the Mode of Existence of Technical Objects*, trans. Cecile Malaspina and John Rogove (Minneapolis: University of Minnesota Press, 2017), 218.

29 ▸ Archimedes, *The Works of Archimedes*, trans. Sir Thomas Heath (Mineola, NY: Dover, 2002), 232.

30 ▸ Aristotle, *The Metaphysics* (London: Penguin Classics, 1999), 18.

31 ▸ For Plato, all of the Forms have the mathematical character of perfect definition in abstract space, even if some, such as "Beauty" or "Justice," are not mathematically derived. Plato, "Timaeus," in *Plato: Complete Works*, ed. John M Cooper and D. S Hutchinson, trans. Donald J. Zeyl (Indianapolis: Hackett, 1997), 1224–91; Plato, *The Republic*, trans. Desmond Lee, 2nd ed. (London: Penguin Classics, 2003).

32 ▸ Benoit B. Mandelbrot, *The Fractal Geometry of Nature* (W. H. Freeman, 1982), 1.

33 ▸ Of course, few workaday mathematicians grapple with such metaphysical consider-ations. Mathematics is generally deployed as a means to derive a practical result or keep an algorithm or other system running. The fantasies of transcendence evoked here lie in the background, on a continuum with other scalar ideologies examined in these pages. My reading of a few foundational mathematical texts in the preceding passage should not be taken as a critique of all mathematics, without which our trans-scalar encounters would be far fewer in number and richness.

34 ▸ Braudel's work overlays multiple temporal and spatial scales in its narration of his-tory. The Annales School's emphasis on historiographical scale led not only to expansions

but also to contractions of scale. One offshoot, dubbed "microhistory," began to focus on the anomalous event, subaltern classes, and histories of singular places. For a discussion of the genealogy and theoretical concerns of microhistory, see, for example, Carlo Ginzburg, "Microhistory: Two or Three Things That I Know about It," trans. John Tedeschi and Anne C. Tedeschi, *Critical Inquiry* 20, no. 1 (October 1993): 10–35. Fernand Braudel, *The Mediterranean and the Mediterranean World in the Age of Philip II*, vol. 1 (Berkeley: University of California Press, 1996).

35 ▸ Manuel De Landa, *A Thousand Years of Nonlinear History* (New York: Zone, 2000).

36 ▸ Michel Foucault, *The Birth of the Clinic: An Archaeology of Medical Perception* (New York: Vintage, 1994); Michel Foucault, *History of Madness*, ed. Jean Khalfa, trans. Jonathan Murphy (New York: Routledge, 2006); Michel Foucault, *Discipline and Punish: The Birth of the Prison* (New York: Vintage, 1995); Henri Lefebvre and Donald Nicholson-Smith, *The Production of Space* (Malden, MA: Blackwell, 2011).

37 ▸ Franco Moretti, *Graphs, Maps, Trees: Abstract Models for Literary History* (New York: Verso, 2007) (distant reading); Matthew L. Jockers, *Macroanalysis: Digital Methods and Literary History* (Urbana: University of Illinois Press, 2013).

38 ▸ See Levi R. Bryant, *The Democracy of Objects* (Ann Arbor: MPublishing, University of Michigan Library, 2011).

39 ▸ Timothy Morton, *Hyperobjects: Philosophy and Ecology after the End of the World* (Minneapolis: University of Minnesota Press, 2013).

40 ▸ See Jane Bennett, *Vibrant Matter: A Political Ecology of Things* (Durham, NC: Duke University Press, 2010); Barad, *Meeting the Universe Halfway*.

41 ▸ Psychologist James J. Gibson coined the term "affordance" to describe the way features of a perceiver's environment offer potential actions to that subject. Media theorist Matthew Fuller influentially extends the concept to apply to mediating systems more generally. In chapter 5, I offer a reading of ecological scalar mediation that draws upon both accounts but primarily deploys Gibson's original reading in the service of theorizing scalar mediation as a set of "resolving cuts" that reveal particular surface details of the environment as a process of milieu-building. See James J. Gibson, *The Ecological Approach to Visual Perception* (New York: Psychology Press, 1986), 45–47.

42 ▸ This is Timothy Morton's useful phrase for others whose strangeness is itself strange to us, that is, whose effects cannot be contained within the identities we prepare for them. See Timothy Morton, "Thinking Ecology: The Mesh, the Strange Stranger, and the Beautiful Soul," *Collapse* 6 (2010): 268.

43 ▸ Marshall McLuhan, *Understanding Media: The Extensions of Man* (New York: McGraw-Hill, 1966), 19.

44 ▸ See Zachary Horton, "The Trans-Scalar Challenge of Ecology," *Interdisciplinary Studies in Literature and Environment* 26, no. 1 (Winter 2019): 5–26.

45 ▸ Félix Guattari, *Chaosmosis: An Ethico-Aesthetic Paradigm*, trans. Julian Pefanis (Bloomington: Indiana University Press, 1995), 33; Matthew Fuller, *Media Ecologies: Materialist Energies in Art and Technoculture* (Cambridge, MA: MIT Press, 2005).

46 ▸ Sarah Kember and Joanna Zylinska, *Life after New Media: Mediation as a Vital Process* (Cambridge, MA: MIT Press, 2012), 22.

47 ▸ Jussi Parikka, *A Geology of Media* (Minneapolis: University of Minnesota Press, 2015), viii, 44.

48 ▸ Parikka, *Geology of Media*, 4.

49 ▸ John Durham Peters, *The Marvelous Clouds: Toward a Philosophy of Elemental Media* (Chicago: University of Chicago Press, 2015), 2, 3.

50 ▸ Peters, *Marvelous Clouds*, 109.

51 ▸ Peters, *Marvelous Clouds*, 51.

52 ▸ Ernst Haeckel, *Generelle Morphologie der Organisme* (Berlin: Reimer, 1866), 2:286; as quoted and translated by Robert C. Stauffer, "Haeckel, Darwin, and Ecology," *Quarterly Review of Biology* 32, no. 2 (June 1957): 140.

53 ▸ Alexander R. Galloway and Eugene Thacker, *The Exploit: A Theory of Networks* (Minneapolis: University of Minnesota Press, 2007), 156.

54 ▸ N. Katherine Hayles, *My Mother Was a Computer: Digital Subjects and Literary Texts* (Chicago: University of Chicago Press, 2005), 7.

55 ▸ Charles Eames and Ray Eames, *A Rough Sketch for a Proposed Film Dealing with the Powers of Ten and the Relative Size of Things in the Universe*, short film (1968).

56 ▸ T. S. Eliot, *The Waste Land and Other Writings* (New York: Modern Library, 2002), 51.

CHAPTER 2

1 ▸ Kees Boeke, letter to James Walsh, July 5, 1956, International Institute of Social History, Amsterdam.

2 ▸ Kees Boeke, *Cosmic View: The Universe in Forty Jumps* (New York: John Day Co., 1957), 7.

3 ▸ Fiona Joseph, *Beatrice: The Cadbury Heiress Who Gave Away Her Fortune* (Birmingham: Foxwell Press, 2012), prologue.

4 ▸ Kees Boeke, "Sociocracy," accessed August 31, 2014, http://worldteacher.faithweb.com/sociocracy.htm.

5 ▸ Kees Boeke, "Bilthoven, Holland's International Children's Community," *Clearing House* 13, no. 2 (1938): 106. The Werkplaats remains in operation to this day.

6 ▸ Boeke, "Bilthoven," 107.

7 ▸ Boeke, "Sociocracy."

8 ▸ Boeke, "Sociocracy."

9 ▸ Boeke, *Cosmic View*, 7.

10 ▸ Boeke, *Cosmic View*, 10.

11 ▸ Boeke, *Cosmic View*, 11.

12 ▸ I am informed by Jos Heuer, the current director of Werkplaats Kindergemeenschap, that the building in question was built and maintained by the Nazi occupation forces as a radio transmission station. As we will see, radio transmissions between Bilthoven and Utrecht become part of *Cosmic View*'s narrative. In an odd way, the Werkplaats as communicative transmission site, as scalar transducer, while prefigured and literalized by the Nazis, is appropriated and fully exploited by Boeke, who renders the site both the source and subject of his culminating experiment in scalar pedagogy, transmitted to the world in codex form.

13 ▸ Boeke, *Cosmic View*, 12.

14 ▸ Boeke, *Cosmic View*, 12.

15 ▸ *Oxford English Dictionary Online*, s.v. "resolution, n.1."

16 ▸ The resolution of the ink-and-paper assemblage was of tantamount importance to the publisher of *Cosmic View*. Publisher Richard Walsh asked the production department for multiple tests of different paper compositions to measure the behavior of the ink droplets. As a result of these tests, the publisher decided to print the book on fine-quality offset, rather than coated, paper. See James Walsh, letter to Kees Boeke, April 11, 1957, International Institute of Social History, Amsterdam.

17 ▸ The degree to which the "original" assemblage or scene conforms to the surfaced version evoked through the medial process—its mimetic authenticity—is not my concern here. For example, Boeke's scale 3 has at least two possible reference points: the illustration on page 11 of *Cosmic View* (the first surface) and the narrative scene involving the U-shaped building, girl, whale, and cat (the second surface). The latter must become a planar surface in order to correspond to the original image. Whether there is a third assemblage, in a particular place and time, and whether we consider it a surface or a volume, is inconsequential. In this case, there never was a conjoined time and place in which these elements were coterminous. They form a narrative surface at a particular scale, in a determinate relationship with the page that evokes them.

18 ▸ Barad, *Meeting the Universe Halfway*, 140.

19 ▸ Barad, *Meeting the Universe Halfway*, 140.

20 ▸ Barad, *Meeting the Universe Halfway*, 149.

21 ▸ Jacques Derrida, *Of Grammatology* (Baltimore: Johns Hopkins University Press, 1998), 158.

22 ▸ Fuller, *Media Ecologies*, 70.

23 ▸ McLuhan, *Understanding Media*, 19.

24 ▸ Gilles Deleuze and Félix Guattari, *Anti-Oedipus: Capitalism and Schizophrenia* (Penguin Classics, 2009), 5.

25 ▸ De Landa, *Thousand Years*, 55.

26 ▸ Nicole Starosielski, *The Undersea Network* (Durham, NC: Duke University Press, 2015), 6.

27 ▸ See Parikka, *Geology of Media*, viii.

28 ▸ For a discussion of the problems introduced by ecology's scale-defined boundaries and a proposal for a conception of ecology that is radically trans-scalar, see Zachary Horton, "Toward a Speculative Nano-Ecology: Trans-Scalar Knowledge, Disciplinary Boundaries, and Ecology's Posthuman Horizon," *Resilience* 2, no. 3 (Fall 2015): 58–86.

29 ▸ Boeke, *Cosmic View*, 13.

30 ▸ Starosielski, *Undersea Network*, 10.

31 ▸ Jakob von Uexküll, "An Introduction to Umwelt," *Semiotica* 134, no. 1/4 (2001): 108.

32 ▸ Boeke, *Cosmic View*, 38.

33 ▸ Ray and Charles Eames label the zooming-out phase of their films the "outward journey" and the zooming-in portion the "inward journey." For the sake of consistency, I've adopted this terminology, not used by Boeke, for all cosmic-zoom manifestations.

34 ▸ Boeke, *Cosmic View*, 9, 11.

35 ▸ Roland Barthes, *Camera Lucida: Reflections on Photography*, trans. Richard Howard (New York: Hill and Wang, 2010), 5.

36 ▸ Boeke, *Cosmic View*, 25.

37 ▸ Boeke, *Cosmic View*, 45.

38 ▸ For an excellent discussion of the interpretative steps in this process, see Jim Gimzewski and Victoria Vesna, "The Nanomeme Syndrome: Blurring of Fact and Fiction in the Construction of a New Science," *Technoetic Arts: A Journal of Speculative Research* 1, no. 1 (January 2003): 7–24.

39 ▸ Boeke, *Cosmic View*, 7.

40 ▸ Boeke further notes that the atom is not the smallest entity in the book, and infinity makes no appearance at all. See Kees Boeke, letter to James Walsh, June 21, 1957, International Institute of Social History, Amsterdam.

CHAPTER 3

1 ▸ Stewart Brand, ed., *Whole Earth Catalog*, vol. 1 (1968), 5.

2 ▸ See Robert Poole, *Earthrise: How Man First Saw the Earth* (New Haven, CT: Yale University Press, 2008), for a history of the cultural forces that anticipate and make use of NASA's earth imagery to further these movements. See Andrew G. Kirk, *Counterculture Green: The Whole Earth Catalog and American Environmentalism* (Lawrence: University Press of Kansas, 2007), for a history of Brand's agitation on behalf of the fusion of ecology, high technology, and systems thinking.

3 ▸ Eames Demetrios, *An Eames Primer* (New York: Universe, 2002), 247.

4 ▸ Michael Golec, "Optical Constancy, Discontinuity, and Nondiscontinuity in the Eameses' Rough Sketch," in *The Educated Eye: Visual Culture and Pedagogy in the Life Sciences*, ed. Nancy Anderson and Michael R. Dietrich (Hanover, NH: Dartmouth College Press, 2012), 167.

5 ▸ Clile C. Allen, "Optical objective," US patent 696788 A, filed February 25, 1901, issued April 1, 1902, http://www.google.com/patents/US696788.

6 ▸ Golec, "Optical Constancy," 174.

7 ▸ *Cosmic Zoom* begins with a live-action scene filmed with a zoom lens, which zooms from an Ottawa shoreline to a boy and a dog in a canoe. The image then fades to animation, and the same "zoom" is made in reverse, on an animation stand. The second, animated version, is indeed flatter than the image produced by the initial lens.

8 ▸ In *Cosmic Zoom*, a bizarre exception proves the rule. When the FOV has expanded to the point that earth's moon becomes visible, it is rendered as having 4.5 times the diameter of the earth! As it emerges from the bottom right corner of the screen, most viewers will read the size difference as a depth cue, suggesting for a brief moment that we are traveling away from the earth and have just whizzed by the surface of the moon. Such a dramatic deviation from the visual modality of the film as a whole (and Boeke's work upon which it is based) is attributable, I think, to the appearance earlier that year of the famous "Earthrise" photograph, taken by Bill Anders aboard Apollo 8. "Earthrise" depicts the edge of the moon in dramatic close-up, with the much smaller earth floating just above. Szasz's rendering of the moon emulates this encounter. The effect of motion is quickly negated, however, as the continual shrinking of the two-dimensional artworks depicting the earth and moon fail to register any change in relative distance, and the earth disappears far sooner than the

moon, an obvious error of perspective. Thus *Cosmic View* belies its own lack of movement by rendering one scale with a depth perspective that could only hold for one instant in time; the time-based nature of the filmic medium immediately gives the lie to this illusion of movement.

9 ▸ Russell Schweickart, "No Frames, No Boundaries," in *Earth's Answer: Explorations of Planetary Culture at the Lindisfarne Conferences*, ed. Michael Katz (Lindisfarne Books/Harper and Row, 1977), 12.

10 ▸ Sylvie Bissonnette, "Scalar Travel Documentaries: Animating the Limits of the Body and Life," *Animation* 9, no. 2 (2014): 139, https://doi.org/10.1177/1746847714526664.

11 ▸ Bissonnette, "Scalar Travel Documentaries," 144.

12 ▸ Denis Cosgrove, *Apollo's Eye: A Cartographic Genealogy of the Earth in the Western Imagination* (Baltimore: Johns Hopkins University Press, 2003), 20, 258.

13 ▸ Bissonnette, "Scalar Travel Documentaries," 154.

14 ▸ Golec, "Optical Constancy," 173, 179, 175.

15 ▸ Gerard 't Hooft and Stefan Vandoren, *Time in Powers of Ten: Natural Phenomena and Their Timescales* (World Scientific Publishing, 2014), xii.

16 ▸ Boeke, *Cosmic View*, 16.

17 ▸ Boeke, *Cosmic View*, 24.

18 ▸ "Powers of Ten: Dashboard" (section of untitled, undated document), Eames Collection, Library of Congress, 1–2.

19 ▸ Golec, "Optical Constancy," 167.

20 ▸ "Red Pencil Narration Draft," n.d., Eames Collection, Library of Congress.

21 ▸ Philip Morrison, "Powers of Ten Script," September 5, 1977, Eames Collection, Library of Congress (underline in original).

22 ▸ Suture, as elaborated by Daniel Dayan, is a function of a film's signifying elements, which together position the viewer within an economy of subjectivization, leading her to identify with certain subject positions constructed by the film (usually the film's protagonist). Laura Mulvey extends the concept as "the gaze," arguing that suture is often gendered, leading the viewer to identify with the subject position of male characters. See Daniel Dayan, "The Tutor-Code of Classical Cinema," *Film Quarterly* 28, no. 1 (Autumn 1974): 22–31; Laura Mulvey, "Visual Pleasure and Narrative Cinema," *Screen* 16, no. 3 (1975): 6–18, https://doi.org/10.1093/screen/16.3.6.

23 ▸ Sam Elton, *A New Model of the Solar System* (New York: Philosophical Library, 1966), quoted in untitled memo.

24 ▸ Elton leaves unspecified exactly who has "found" this or what means were employed to do so. See Elton, *New Model*, 116.

25 ▸ For a discussion of the microcosmic theories of Paracelsus as a comparative model of scalar relationships, see Zachary Horton, "Composing a Cosmic View: Three Alternatives for Thinking Scale in the Anthropocene," in *Scale in Literature and Culture*, ed. Michael Travel Clarke and David Wittenberg (New York: Palgrave Macmillan, 2017), 35–60.

26 ▸ Paracelsus, "Hermetic Astronomy," in *The Hermetic and Alchemical Writings of Paracelsus*, ed. Arthur Edward Waite (Martino Fine Books, 2009), 2:285.

27 ▸ Schrader, in identifying only Charles as the author of the Eameses' films, unfortunately perpetuates the sexist assumption, a vestige of the cultural milieu in which the Eameses rose to fame, that Charles was the primary creative force behind their collaborative work. In the decades following Charles's death in 1977 this misconception was progressively put to rest. Ray and Charles collaborated in nearly everything they did, producing a domestic-creative-business symbiot that functioned 24-7 for several decades. Nonetheless, Charles was more interested than Ray in photography and filmmaking, and he does indeed seem to have been the primary *directorial* force behind their films. Ray, however, was heavily involved in the *aesthetic design* of their films, and I will continue to refer to them as jointly authored. For a revealing exploration of the Eameses' collaborative and domestic dynamics and the personal frustration and embarrassment they experienced as a result of their social milieu's asymmetrical and gendered assignation of credit, see Jason Cohn and Bill Jersey's 2012 documentary *Eames: The Architect & The Painter*.

28 ▸ Charles Eames and Ray Eames, *Blacktop: A Story of the Washing of a School Play Yard* (1952) and *Day of the Dead* (1957).

29 ▸ Paul Schrader, "Poetry of Ideas: The Films of Charles Eames," *Film Quarterly* 23, no. 3 (1970): 2.

30 ▸ Eric Schuldenfrei, *The Films of Charles and Ray Eames: A Universal Sense of Expectation* (Routledge, 2015), 114.

31 ▸ Hans Christian Von Baeyer, "Power Tool," *Sciences* 40, no. 5 (2000): 15, https://doi .org/10.1002/j.2326-1951.2000.tb03516.x.

32 ▸ Schrader, "Poetry of Ideas," 7 (emphasis added).

33 ▸ Schrader, "Poetry of Ideas," 12.

34 ▸ Elizabeth Kessler suggests that the Eameses assumed public familiarity with powers when they chose this as their primary numerical display in *Powers of Ten* (which omits most of the rest of *Rough Sketch*'s instrument panel). Familiarity with such notation would have stemmed, Kessler argues, from the ubiquity, at the time, of the slide rule as a calculating device. Slide rules are composed of sliding logarithmic scales. See Elizabeth A. Kessler, "Jumps in Space: The Disappearing and Appearing Logarithmic Scale," Society for Literature, Science, and the Arts, Dallas, TX, 2014.

35 ▸ Philip Morrison and Phylis Morrison, "A Happy Octopus," in *The Work of Charles and Ray Eames: A Legacy of Invention*, ed. Donald Albrecht (New York: Harry N. Abrams, 1997), 107, 106.

36 ▸ John Neuhart, Marilyn Neuhart, and Ray Eames, *Eames Design* (New York: Harry N. Abrams, 1989), 356.

37 ▸ Demetrios, *Eames Primer*, 148.

38 ▸ Charles Eames and Ray Eames, *Toccata for Toy Trains*, short film, 1957.

39 ▸ Charles Eames and Ray Eames, *Design Q&A*, short film, 1972.

40 ▸ Justus Nieland, "Midcentury Futurisms: Expanded Cinema, Design, and the Modernist Sensorium," *Affirmations of the Modern* 2, no. 1 (2015): 80, http://am.ubiquitypress .com/articles/59/.

41 ▸ Philip Morrison, professor of physics at MIT, not only admired *Tops*, but frequently showed the film to students to demonstrate "the applicability of science to all aspects of life." Morrison features prominently in chapter 4. Pat Kirkham, *Charles and Ray Eames: Designers of the Twentieth Century* (Cambridge, MA: MIT Press, 1995), 336.

42 ▸ Stewart, in discussing the "miniature," makes no distinction between the miniature and the toy. In the Eames lexicon I have adopted in this chapter, her discussion better fits the toy. The miniature, for Stewart, "typifies the structure of memory, of childhood, and ultimately of narrative's secondary (and at the same time causal) relation to history." Susan Stewart, *On Longing: Narratives of the Miniature, the Gigantic, the Souvenir, the Collection* (Durham, NC: Duke University Press, 1992), 69, 171.

43 ▸ For the converse—the resolving of the non-toy environment's material affordances—see chapter 5.

44 ▸ Brian Aldiss dramatizes the point the following year in his short story "Super-Toys Last All Summer Long," which figures a manufactured human simulacrum that substitutes for a biological child in a population-controlled future. The android's "mother" rejects the child, who is left to ponder, along with its AI-driven teddy bear, "Are we real?" Stanley Kubrick had the story adapted into a screenplay for *A.I.*, a film Steven Spielberg ultimately directed after Kubrick's death. See Aldiss, "Super-Toys Last All Summer Long," *Harper's Bazaar*, December 1969.

45 ▸ Fred Turner, *The Democratic Surround: Multimedia and American Liberalism from World War II to the Psychedelic Sixties* (Chicago: University of Chicago Press, 2013), 280.

46 ▸ *Think* was thirty minutes in length and projected upon twenty-two screens in an elevated, ovoid theater constructed for the IBM Pavilion in Queens, New York. The images related to a series of problems from different scales and suggested how they could be solved through information processing. "Sequences in the film dealt with problems of varying complexity, ranging from common, everyday ones—how to plan the seating for a dinner party—to complex business and scientific problems—city planning and weather prediction." Neuhart, Neuhart, and Eames, *Eames Design*, 289.

CHAPTER 4

1 ▸ Demetrios, *Eames Primer*, 249.

2 ▸ Demetrios, *Eames Primer*, 249.

3 ▸ Geoffrey A. Moore, letter to Ann Jones, undated, Library of Congress.

4 ▸ Work was delayed for a time by a legal dispute. The American publishing firm John Day held worldwide distribution rights to Kees Boeke's *Cosmic View* and thus any film based upon the text. It had licensed the material to Atlantis Productions in 1957 for a little-seen and completely unavailable (even by 1971) educational film titled *How Vast Is Space?* Atlantis Productions sued the Eames Office in 1971 for copyright infringement. See Office of Johnson and Tannenbaum, letter to Offices of Kaplan, Livingston, Goodwin, Berkowitz & Selvin, August 13, 1971, Eames Collection, Library of Congress. The Eameses settled out of court, agreeing to license the film rights and pay royalties on every copy of *Rough Sketch* sold. Any subsequent film would, however, be free of such royalty obligations. See Leonard Alexander, letter to Gary Concoff, June 17, 1972.

5 ▸ These letters, and the Eames Office's replies, are preserved in the Eames Collection, in the Library of Congress.

6 ▸ Funke would go on to head the special-effects department for Peter Jackson's *Lord of the Rings* trilogy.

7 ▸ "Annotated *Rough Sketch* Storyboard," n.d., Eames Collection, Library of Congress.

8 ▸ Morrison, "Powers of Ten Script," September 5, 1977, 5.

9 ▸ The Eameses produced a short film on the topic, *Topology*, in 1962.

10 ▸ Philip Morrison, "Something Personal" section introduction," in *Nothing Is Too Wonderful to Be True* (Woodbury, NY: American Institute of Physics, 1997), 2.

11 ▸ A. P. French, "In Memoriam," *Physics in Perspective* 10, no. 1 (2008): 113–14, https://doi.org/10.1007/s00016-007-0343-5.

12 ▸ Philip Morrison, "If the Bomb Gets Out of Hand," in *Nothing Is Too Wonderful to Be True* (Woodbury, NY: American Institute of Physics, 1997), 291–99.

13 ▸ Philip Morrison, "The Actuary of Our Species," in *Nothing Is Too Wonderful to Be True* (Woodbury, NY: American Institute of Physics, 1997), 86.

14 ▸ Philip Morrison, "Life beyond Earth and the Mind of Man," in *Nothing Is Too Wonderful to Be True* (Woodbury, NY: American Institute of Physics, 1997), 175.

15 ▸ For an account of Morrison's role in starting the precursor to the SETI program, see Philip Morrison, "A Talk with Philip Morrison," in *Nothing Is Too Wonderful to Be True* (Woodbury, NY: American Institute of Physics, 1997), 196–203.

16 ▸ Philip Morrison, "The Wonder of Time," in *Nothing Is Too Wonderful to Be True* (Woodbury, NY: American Institute of Physics, 1997), 23.

17 ▸ Philip Morrison, "Knowing Where You Are," in *Nothing Is Too Wonderful to Be True* (Woodbury, NY: American Institute of Physics, 1997), 287.

18 ▸ Morrison, "Wonder of Time," 28.

19 ▸ French, "In Memoriam," 117.

20 ▸ Philip Morrison, "Powers of Ten Script," September 12, 1977, Eames Collection, Library of Congress, 6.

21 ▸ Morrison, "Powers of Ten Script," September 12, 1977, 6.

22 ▸ Morrison, "Powers of Ten Script," September 5, 1977, 7.

23 ▸ Morrison, "Powers of Ten Script," September 5, 1977, 8.

24 ▸ Morrison, "Powers of Ten Script," September 12, 1977, 8.

25 ▸ Theophrastus, "De Causis Plantarum," in *Selections from Early Greek Philosophy*, ed. Milton C. Nahm, 4th ed. (New York: Appleton-Century-Crofts, 1964), 177.

26 ▸ Morrison, "Wonder of Time," 19, 23.

27 ▸ This point was important enough for the Eames Office to explicitly emphasize it to the French translators of the narration. "Important to end with the forty zeros, as you recognized," notes a letter dated April 24, 1979. See Jehane Burns, letter to Brian Southworth, April 24, 1979, Eames Collection, Library of Congress.

28 ▸ For a visual comparison of many of the works discussed here, see chapter 5, figure 25.

29 ▸ Nonetheless, as I have previously noted, *Cosmic View* is rich in scalar intra-actions— events or dynamics taking place within particular scales: 90.2 percent of Boeke's scales contain intrascalar dynamics, compared with only 56.1 percent of *Powers of Ten*'s scales (these are represented as shaded nodes in my network graphs).

30 ▸ Viewing context can of course coax forth one reading rather than another. When shown in a darkened theater or classroom, *Powers of Ten*'s mesmeric continuous zoom tends to produce an overwhelming sense of scalar contiguity. Repeated viewings may,

however, attenuate the mesmeric quality and help to bring out the network reading. The film was and is also frequently screened in science museums, where the discretized, disciplinary nature of the exhibits may cue viewers to make similar resolving cuts in the film's otherwise unbroken visual stream.

31 ▸ See Ursula K. Heise, *Sense of Place and Sense of Planet: The Environmental Imagination of the Global* (New York: Oxford University Press, 2008), esp. chap. 2.

32 ▸ "Dashboard Display for Revised Format," n.d., Eames Collection, Library of Congress, 2.

33 ▸ "Untitled Dashboard Mockup," n.d., Eames Collection, Library of Congress. Phylis's direct contributions to the film seem to have been primarily related to research into scanning electron microscopy, but Eames Office references are usually to "the Morrisons," plural, so it is reasonable to assume that Philip consulted with her about many of his production decisions.

34 ▸ Neuhart, Neuhart, and Eames, *Eames Design*, 441.

35 ▸ "Powers of Ten Exploration," n.d., Eames Collection, Library of Congress, 1.

36 ▸ Schuldenfrei, *Films of Charles and Ray Eames*, 137.

37 ▸ Neuhart, Neuhart, and Eames, *Eames Design*, 441.

38 ▸ Bissonnette, "Scalar Travel Documentaries," 145.

39 ▸ "Powers of Ten Brochure," 1977, Eames Collection, Library of Congress.

40 ▸ I use the gendered noun here quite deliberately.

41 ▸ Derek Woods, "Epistemic Things in Charles and Ray Eames's Powers of Ten," in *Scale in Literature and Culture*, ed. Michael Travel Clarke and David Wittenberg (New York: Palgrave Macmillan, 2017), 86.

42 ▸ J. A. Wiens, "Spatial Scaling in Ecology," *Functional Ecology* 3, no. 4 (1989): 385–97.

43 ▸ Horton, "Trans-Scalar Challenge."

44 ▸ Cosgrove, *Apollo's Eye*, xi.

45 ▸ Boeke, *Cosmic View*, 35

46 ▸ Boeke's first draft of this text began with the following line: "More important is the fact that '#27' would in any case be only partly filled, as what is called the end of 'the ray of the universe' would fit into the scale of this drawing." His editor, Richard Walsh, was sufficiently confused by this assertion to write to Boeke asking what "the ray of the universe" meant. Boeke responded by omitting it from the next draft of the text. This perhaps suggests that textual description is also largely inadequate to capture the dynamics of this largest of scales. Nonetheless, Boeke here reminds the reader once again that a given scale is not merely an expanded view arrayed contiguously with adjacent scales by problematizing the optical register's ability to properly capture scalar difference. See James Walsh, letter to Kees Boeke, May 16, 1957, International Institute of Social History, Amsterdam.

47 ▸ Eames and Eames, *Rough Sketch*.

48 ▸ Martin Heidegger, "The Age of the World Picture," in *The Question Concerning Technology, and Other Essays* (New York: Harper Torchbooks, 1977), 130, 128.

49 ▸ Despite this insight, Heidegger fails to account for this cultural transmission process or, in Bruno Latour's terms, how social structures become stabilized as persisting assem-

blages through the agency of "a world made of *concatenations of mediators* where each point can be said to fully act." See Bruno Latour, *Reassembling the Social: An Introduction to Actor-Network-Theory* (New York: Oxford University Press, 2007), 59.

50 ▸ I use "alterior" as an adjective form of "alterity," or ontological difference. I mean the term to imply not just difference in general, but the difference that lies at the bottom of any trans-scalar encounter; that is, the difference that either irrupts consciousness or is collapsed through trans-scalar media.

51 ▸ Schuldenfrei, *Films of Charles and Ray Eames*, 75.

52 ▸ Neuhart, Neuhart, and Eames, *Eames Design*, 239–40.

53 ▸ Val Plumwood, *Feminism and the Mastery of Nature* (London: Routledge, 1994), 20.

54 ▸ Bruno Latour, "Anti-Zoom," in *Scale in Literature and Culture*, ed. Michael Travel Clarke and David Wittenberg (New York: Palgrave Macmillan, 2017), 94.

55 ▸ Curiously, the *Powers of Ten* book fails to describe intrascalar interactions at the scale of the earth and the earth-moon system. *Cosmic View*, by contrast, omits interactions within the atom.

56 ▸ I have used the same layout engine, Force Atlas 2 (in the software package Gephi), to generate all of the network graphs in this book, in order to facilitate comparisons. I manually moved certain nodes slightly to minimize overlap and increase legibility. That the *POT* book data would produce the semblance of a fish (figure 20) is accidental, an algorithmic fluke. Or is it? I have endeavored to describe this network structure in terms of its connective affordances or potential path structures. Any work of scalar mediation complex enough to aggregate many scales will produce such a set of affordances, giving rise to a distinctive shape that is, in a sense, a virtual scalar organism. This brings to mind Steven Hall's magnificent novel *The Raw Shark Texts*, which imagines sharks and other quasi-marine creatures evolving within the vast ocean of data produced by the human race, creatures that live in their own virtual ecosystem, feeding off each other. The scalar fish of figure 20 may be accidental, but are we not all accidental fish swimming in the cosmic ocean? See Steven Hall, *The Raw Shark Texts: A Novel* (Canongate U.S., 2008).

57 ▸ Howard Boyer, letter to Ray Eames, August 27, 1981, Eames Collection, Library of Congress.

58 ▸ Charles Eames and Ray Eames, *Powers of Ten: A Flipbook* (New York: W. H. Freeman, 1998).

59 ▸ Philip Morrison, "Looking at the World: An Essay," in Philip Morrison, Phylis Morrison, and Ray Eames, *Powers of Ten: About the Relative Size of Things in the Universe* (Redding, CT: Scientific American Library, 1982), 4.

60 ▸ Bruno Latour, "Visualisation and Cognition: Drawings Things Together," *Knowledge and Society: Studies in the Sociology of Culture Past and Present* 6 (1986): 13, 19.

61 ▸ Morrison, "Looking at the World," 3.

62 ▸ Neuhart, Neuhart, and Eames, *Eames Design*, 405–10.

63 ▸ Isaac Newton, *The Principia: Mathematical Principles of Natural Philosophy*, trans. Andrew Motte, 3rd ed. (CreateSpace, 2013), 320.

64 ▸ Isaac Newton, *A Historical Account of Two Notable Corruptions of Scripture* (Rough Draft Printing, 2011).

65 ▸ Newton, *Historical Account*, 440.

66 ▸ Newton, *Principia*, 442. Spinoza's famous argument is that God cannot exist separate from the universe and thus could not have created it as something outside of himself. "Outside of God there can be no substance, that is, thing which is in itself outside God. . . . God, therefore, is the immanent, not the transitive cause of all things." Many commentators have understood this to entail that God and nature are the same thing; in other words, the universe *just is* God. See Benedict de Spinoza, *Ethics*, trans. Edwin Curley (London: Penguin Classics, 2005), 16.

67 ▸ Newton, *Principia*, 440.

68 ▸ Isaac Newton, *Observations upon the Prophecies of Daniel and the Apocalypse of St. John* (Watchmaker Publishing, 2011), 162.

69 ▸ Neuhart, Neuhart, and Eames, *Eames Design*, 405.

70 ▸ Morrison, "Wonder of Time," 19.

71 ▸ Philip Morrison, "Cause, Chance, and Creation," in *Nothing Is Too Wonderful to Be True* (Woodbury, NY: American Institute of Physics, 1997), 88, 89.

72 ▸ Morrison, "Looking at the World," 8.

73 ▸ Morrison, "Looking at the World," 6.

74 ▸ Philip Morrison and Phylis Morrison, preface, in Morrison, Morrison, and Eames, *Powers of Ten*.

75 ▸ Morrison, "Looking at the World," 8.

76 ▸ Morrison, "Looking at the World," 8.

77 ▸ Ian Hacking, *Representing and Intervening* (Cambridge: Cambridge University Press, 1983), 31.

78 ▸ Morrison, "Looking at the World," 9.

79 ▸ Morrison, "Looking at the World," 6.

80 ▸ Latour, *Reassembling the Social*, 83.

81 ▸ In the *Powers of Ten* film, as I've noted, the Eameses showed the whole earth within the blue scalar square nominally indicating 10^7 meters, though the planet is actually larger; in the book, as in *Rough Draft*, the planet and blue square are accurately scaled.

82 ▸ Caleb Scharf, *The Zoomable Universe: An Epic Tour Through Cosmic Scale, from Almost Everything to Nearly Nothing* (New York: Farrar, Straus and Giroux, 2017), 152.

83 ▸ Scharf, *Zoomable Universe*, 148.

84 ▸ Chapter 6 will discuss the relationship between digital media and scale more explicitly.

85 ▸ Scharf, *Zoomable Universe*, 93.

86 ▸ Gabrielle Hecht, "Interscalar Vehicles for an African Anthropocene: On Waste, Temporality, and Violence," *Cultural Anthropology* 33, no. 1 (2018): 114, https://doi.org/10.14506/ca33.1.05.

CHAPTER 5

1 ▸ Michael D. Uchic et al., "Sample Dimensions Influence Strength and Crystal Plasticity," *Science* 305, no. 5686 (2004): 986, https://doi.org/10.1126/science.1098993.

2 ▸ Bonner, *Why Size Matters*, 3–4.

3 ▸ Bonner, *Why Size Matters*, 5.

4 ▸ Thompson, *On Growth and Form*, 17.

5 ▸ Andrew Pickering, *The Mangle of Practice: Time, Agency, and Science* (Chicago: University of Chicago Press, 1995), 116, 115.

6 ▸ Pickering, *Mangle of Practice*, 129, 131.

7 ▸ Harry M. Collins, *Changing Order: Replication and Induction in Scientific Practice* (Chicago: University of Chicago Press, 1992), 84.

8 ▸ Collins, *Changing Order*, 144–45.

9 ▸ A good example is elaborated by Collins and Pinch in *The Golem*: the infamous cold fusion research by Martin Fleishmann and Stanley Pons. These scientists hailed from the field of chemistry, and their experiment was formulated according to the protocols of that discipline, but their extraordinary claim at a press conference in 1989 was that they had discovered a new energy source in the form of nuclear fusion. This was the disciplinary terrain of physics, a discipline focused not at the scale of chemical solutions and exothermic reactions but at the atomic scale. While chemists were initially persuaded by the claims, primarily due to the reaction's exothermic (heat-releasing) result, physicists were vocally unconvinced that Fleishmann and Pons could make any claims at the molecular scale—to wit, that deuterium molecules were fusing nuclei from hydrogen. Their arguments won in the larger social field of scientific inquiry, and Fleishmann and Pons were unable to obtain further funding, despite the fact that neither their chemistry-scaled experiment nor their interpretation of the result was ever disconfirmed. Harry Collins and Trevor Pinch, *The Golem: What Everyone Should Know about Science* (Cambridge: Cambridge University Press, 1993), 76–77.

10 ▸ Bennett, *Vibrant Matter*, xvi; Guattari, *Chaosmosis*, 40; Jay McDaniel, "Physical Matter as Creative and Sentient," ed. Eugene C. Hargrove, *Environmental Ethics* 5, no. 4 (1983): 292, https://doi.org/10.5840/enviroethics19835413. Guattari views machines as prior to technology and independent of the human (though humans form and are formed of machines at many scales). Machines, in this view, are autopoeitic systems subject to evolution; thus, we must condition ourselves to think "a mecanosphere superposed on the biosphere."

11 ▸ Lucretius, *The Nature of Things*, trans. Alicia Stallings (London: Penguin Classics, 2007), 42.

12 ▸ Italo Calvino, "All at One Point," in *Cosmicomics*, trans. William Weaver (New York: Harcourt Brace Jovanovich, 1976), 44.

13 ▸ Gilles Deleuze, *Difference and Repetition*, trans. Paul Patton (New York: Columbia University Press, 1995), 40–41.

14 ▸ Deleuze, *Difference and Repetition*, 262.

15 ▸ Deleuze, *Difference and Repetition*, 222.

16 ▸ Deleuze, *Difference and Repetition*, 233.

17 ▸ Deleuze, *Difference and Repetition*, 266.

18 ▸ Jacques Derrida, "Différance," in *Speech and Phenomena and Other Essays on Husserl's Theory of Signs*, trans. David B. Allison (Evanston, IL: Northwestern University Press, 1973), 138.

19 ▸ Jacques Derrida, "Structure, Sign, and Play in the Discourse of the Human Sciences," in *Writing and Difference*, trans. Alan Bass (Chicago: University of Chicago Press, 1980), 369.

20 ▸ Nietzsche's posthumously published essay "On Truth and Lie in an Extra-Moral Sense" begins, "In some remote corner of the universe, poured out and glittering in innumerable solar systems, there once was a star on which clever animals invented knowledge. That was the haughtiest and most mendacious minute of 'world history'—yet only a minute. After nature had drawn a few breaths the star grew cold, and the clever animals had to die." For Nietzsche, human knowledge (including language) is only a tool for living, embedded within an actual milieu. He here invokes the temporal and spatial scale of being to ridicule the pretensions of a species capable of placing its own paltry constructions above being itself. While Nietzsche rejects the idea of objective truth, viewing truths as mere linguistic conventions that have lost their connection to reality; he does so in the service of direct experience and action. Thus, while Derrida borrows Nietzsche's linguistics, he fails to take up its traveling companion, materialism. Nietzschean affirmation fundamentally relies upon the articulation of the two, the production of the generically new as a way to unblock physical forces and effect action in a material milieu. Even acknowledging Derrida's tremendous insight into language, I therefore find that his claim to be an heir of Nietzschean affirmation rings a bit hollow. See Friedrich Nietzsche, "On Truth and Lie in an Extra-Moral Sense," in *The Portable Nietzsche*, trans. Walter Kaufmann (New York: Viking, 1954), 42.

21 ▸ Immanuel Kant, *Critique of Judgment*, trans. Werner S. Pluhar (Indianapolis, IN: Hackett, 1987), 115, 116.

22 ▸ Deleuze, *Difference and Repetition*, 199.

23 ▸ Othmar H. Amman, Theodore von Kármán, and Glenn B. Woodruff, "The Failure of the Tacoma Narrows Bridge," report to the Federal Works Agency, 1941, 113, http://resolver.caltech.edu/CaltechAUTHORS:20140512-105559175.

24 ▸ Gibson, *Ecological Approach*, 127.

25 ▸ Gibson, *Ecological Approach*, 16.

26 ▸ Gibson, *Ecological Approach*, 221.

27 ▸ Gibson, *Ecological Approach*, 255.

28 ▸ Gibson, *Ecological Approach*, 9.

29 ▸ Friedrich Wilhelm Nietzsche, *Beyond Good and Evil*, trans. Marion Faber (Oxford: Oxford University Press, 2008), 14.

30 ▸ Gibson develops his own model of the "picture" to explain how it intervenes in the processes of ecological optics as "both a surface in its own right and a display of information about something else." To generate an image of another surface is not, he emphatically stresses, to copy that surface. "Drawing is never copying. It is impossible to copy a piece of the environment. Only another drawing can be copied." He thus outlines his own nonrepresentational theory of the image that avoids some of the pitfalls of a representational model. In emphasizing media and scale more directly, I have endeavored to offer a model of scalar mediation that more directly accounts for the stabilization of scales and the cultural mediations of technological systems, and have thus developed my own theory of scalar mediation far beyond Gibson's trailblazing insights without substantially deviating from them. Nonetheless, for Gibson, ecological optics are deeply entwined with mobile instruments of visualization. He thus greatly prefers cinematic imagery to still, or "arrested," images because it more closely replicates the process of perceiving an environ-

ment, including the extraction of invariants from the flux of imagistic stimuli. What Gibson does not consider are precisely the processes that I have explored as "scalar mediation." When we consider the mediation of scale as a central process of differentiation, the functions and value of imagistic surfaces take on a significantly greater importance in stabilizing relationships between surfaces that enhance the affordances of their surfaces by greatly extending the potential perception of detail (resolution). From this perspective, the distinction between motion and still imagery becomes less important than the particular details resolved by mediating systems, along with the ways they collapse or retain scalar difference, the precise scalar relationships they stabilize, and the constraints and affordances of trans-scalar access that they enable. See Gibson, *Ecological Approach*, 273, 279.

31 ▸ Fredric Jameson, *Postmodernism; or, The Cultural Logic of Late Capitalism* (Durham, NC: Duke University Press, 1999), 38.

32 ▸ Jean Baudrillard, "The Precession of Simulacra," in *Simulacra and Simulation*, trans. Sheila Faria Glaser (Ann Arbor: University of Michigan Press, 1994), 2.

33 ▸ Eames Office and DATT Japan, *Powers of Ten Interactive*, Macintosh/Windows (Pyramid Media, 1999).

34 ▸ Eames Demetrios, "'Stations through Time' in Digital Anthology, Powers of Ten Interactive CD-ROM" (Pyramid Media, 1999), 1.

35 ▸ We must note, however, that like most highly connected networks, this one is not *perfectly* connected, which would require every node to be connected to every other node, in both directions (as this is a directed graph). A network is said to be perfectly connected if it has a "density" of 1. Our network has a density of 38.8, meaning that almost 39 percent of all possible connections actually exist. This certainly qualifies it as having a structure analogous to a biological rhizome, as well as the diagrammatic (philosophical) rhizome structure described by Deleuze and Guattari. See Gilles Deleuze and Félix Guattari, *A Thousand Plateaus: Capitalism and Schizophrenia* (Minneapolis: University of Minnesota Press, 1987), introduction.

36 ▸ This interface is sufficiently complex and counterintuitive to warrant not one but *two* lengthy tutorial videos, shot by Eames Demetrios after the production of the CD-ROM. The videos were offered as downloads at www.powersof10.com, where they presumably were meant to serve a promotional purpose as well. They are included in their entirety on a later video compilation, *Scale Is the New Geography*. See Eames Demetrios, *Scale Is the New Geography*, DVD, 2008.

37 ▸ On the other hand, while the Eames Office favored intuitive interfaces that receded into the background and provided the most utility for the least amount of work for their users, some of its works did deliberately produce media environments designed to nearly overwhelm the senses (such as the multiscreen film *Glimpses of the USA*) or simply invite unstructured exploration or play. Examples of the latter include *House of Cards*, which invited the user to construct structures out of playing cards depicting various patterns or everyday objects, and some of their non-timelined museum exhibits, such as *Mathematica: A World of Numbers . . . and Beyond* (1961). These works incorporate a rhizomatic structure prefiguring *Powers of Ten Interactive*, albeit without the latter's overwhelming condensation of different interface modalities on a single surface.

38 ▸ *Powers of Ten Interactive* was developed for Windows 95 and what is today referred to as Mac OS Classic. In 2000 Apple released OSX, an entirely new operating-system architecture unable to run most previous software. In 2001 Microsoft released Windows XP, also incompatible with most previous software. Built over five years on its own proprietary platform, the *POT* software could not be ported to newer architectures without being

rebuilt from scratch, and thus quickly become non-executable. Today it can be fully run only on a vintage computer running a vintage operating system, or on a virtual machine running the same. In the latter case, results vary.

CHAPTER 6

1 ▸ Danah Boyd and Kate Crawford, "Critical Questions for Big Data," *Information, Communication & Society* 15, no. 5 (2012): 663, https://doi.org/10.1080/1369118X.2012 .678878.

2 ▸ Elizabeth A. Kessler, *Picturing the Cosmos: Hubble Space Telescope Images and the Astronomical Sublime* (Minneapolis: University of Minnesota Press, 2012), 86.

3 ▸ Lisa Parks, *Cultures in Orbit: Satellites and the Televisual* (Durham, NC: Duke University Press, 2005), 69.

4 ▸ Ursula K. Heise, *Imagining Extinction: The Cultural Meanings of Endangered Species* (Chicago: University of Chicago Press, 2016), 62.

5 ▸ John Cheney-Lippold, *We Are Data: Algorithms and the Making of Our Digital Selves* (New York: NYU Press, 2017), 89.

6 ▸ Graham Harwood, "Pixel," in *Software Studies: A Lexicon*, ed. Matthew Fuller (Cambridge, MA: MIT Press, 2008), 213–17, 215.

7 ▸ James T. Hyder, "The Making of Cosmic Voyage," ed. Maureen Kerr, Jo Hinkel, and Helen Morrill (Educational Services Department, National Air and Space Museum, Smithsonian Institution, 1996), 2.

8 ▸ Hyder, "Making of Cosmic Voyage," 2.

9 ▸ Peter Golkin, "IMAX Audiences Embark on a Cosmic Voyage through Time and Space," National Air and Space Museum, June 24, 1996, http://airandspace.si.edu /about/newsroom/release/?id=111.

10 ▸ While the *Cosmic Voyage* team turned to Hollywood for its subjective and smooth Steadicam aesthetics, Hollywood director Barry Sonnenfeld roundly satirized the film— and the previous Eames versions—a year later in his 1997 blockbuster *Men in Black*. The final shot of the film tracks Will Smith's car driving away, then pulls back in a cosmic zoom of many orders of magnitude. This zoom ends with a twist, however: it continues beyond the edges of human knowledge, revealing the universe to be a bounded marble. A sinewy, tentaclelike hand flicks it against another orb in a personal game of marbles, then drops it into a sack of universes. This, one of my favorite versions of the cosmic zoom, pokes fun at the pretensions of a universal overview of all scales that implicitly places the human at the center of a single homogeneous plane of knowledge.

11 ▸ Vivian Sobchack, "Meta-Morphing," Telepolis, March 18, 1997, https://www.heise.de /tp/features/Meta-Morphing-3445963.html, 43, 44. See also Vivian Sobchack, ed., *Meta-Morphing: Visual Transformation and the Culture of Quick-Change* (Minneapolis: University of Minnesota Press, 2000).

12 ▸ Perhaps not coincidentally, Eames Demetrios published, within one year of *Cosmic Voyage* and Sobchack's article (1997), a new remediation of *Powers of Ten*: a flipbook. After the advent of the pixel-enabled morph, the cosmic zoom had lost its directionality— indeed, had been stripped of all temporality.

13 ▸ Kristen Whissel, "The Digital Multitude," *Cinema Journal* 49, no. 4 (2010): 105, https://doi.org/10.1353/cj.2010.0005.

14 ▸ Lev Manovich, *The Language of New Media* (Cambridge, MA: MIT Press, 2002), 225.

15 ▸ N. Katherine Hayles, *How We Think: Digital Media and Contemporary Technogenesis* (Chicago: University of Chicago Press, 2012), 176.

16 ▸ IBM's monumental Information Management System (IMS) was the first large-scale, commercial hierarchical database system. Developed for NASA's Apollo program, to keep track of its massive bill of materials, it was first deployed in 1968, the same year as the Eameses' *Rough Sketch*. IMS has been in continual development since, and despite the popularity of relational databases today, is still the primary system used in the banking and finance industries worldwide, according to IBM. See IBM Knowledge Center, "History of IMS: Beginnings at NASA," https://www.ibm.com/support/knowledgecenter/zosbasics /com.ibm.imsintro.doc.intro/ip0ind0011003710.htm, and "IMS Is Strategic for Addressing Customer Needs," https://www.ibm.com/support/knowledgecenter/zosbasics/com.ibm .imsintro.doc.intro/ip0inicmst03.htm.

17 ▸ C. J. Date, *An Introduction to Database Systems* (Reading, MA: Addison-Wesley, 1975), 47.

18 ▸ Eugene Thacker, "Networks, Swarms, Multitudes," *CTHEORY*, no. a142 (May 18, 2004), http://www.ctheory.net/articles.aspx?id=422.

19 ▸ E. F. Codd, "A Relational Model of Data for Large Shared Data Banks," *Communications of the ACM* 13, no. 6 (1970): 377–87, https://doi.org/10.1145/362384.362685, 2.

20 ▸ Foucault, *Discipline and Punish*, 139.

21 ▸ David Maier, *The Theory of Relational Databases* (Computer Science Press, 1983), 1.

22 ▸ Date, *Introduction to Database Systems*, 4.

23 ▸ More recently, a fourth type of database, the object- or document-oriented database, has been developed. This sort of database allows every entry to be represented in a different form, defined according to unique attributes—not encoded via relational schema. In many ways, the object-oriented database is a throwback to the hierarchical and network models in that it attempts to encode as much real-world information as possible into its internal structure. As Hayles puts it, "Object-oriented databases encourage a view of the world that is holistic and behavioral, whereas relational databases see the world as made up of atomized bits that can be manipulated through commands." The holistic world modeled by the object-oriented database comes at a cost: documents or objects are not inherently relatable to each other unless they happen to share metadata tags. In other words, these databases fetishize the individual object, not the schema. This tendency toward encoding unique, isolated objects reduces the database's virtual dimension: it becomes less easy to connect one dataset to another, given that they are less recomposable. For these reasons, relational databases remain dominant for most "big data" purposes where data needs to be efficiently sorted and capaciously recomposed. Object- or document-oriented databases (such as MongoDB) are more popular in contexts where many different entities need to be mapped and recalled based on coarse-grained features, such as Internet-connected apps and search engines. See Hayles, *How We Think*, 193.

24 ▸ Robert Kowalski, "Algorithm = Logic + Control," *Communications of the ACM* 22, no. 7 (1979): 429.

25 ▸ Maier, *Theory of Relational Databases*, 224.

26 ▸ Alexander R. Galloway, *The Interface Effect* (Cambridge, UK: Polity, 2012), 52, 54.

27 ▸ Michael Huang and Cary Huang, *The Scale of the Universe 2*, 2012, http://htwins .net/scale2/.

28 ▸ Foucault, *Order of Things*, xvii.

29 ▸ Foucault, *Order of Things*, xvii.

30 ▸ See Geoffrey C. Bowker and Susan Leigh Star, *Sorting Things Out: Classification and Its Consequences*, rev. ed. (Cambridge, MA: MIT Press, 2000), esp. chap. 2.

31 ▸ Spinoza, *Ethics*, 71.

32 ▸ Mark B. N. Hansen, *Feed-Forward: On the Future of Twenty-First-Century Media* (Chicago: University of Chicago Press, 2015), 6.

33 ▸ Cheney-Lippold, *We Are Data*, 9–10.

34 ▸ Seb Franklin, *Control: Digitality as Cultural Logic* (Cambridge, MA: MIT Press, 2015), 11.

35 ▸ Wendy Hui Kyong Chun, *Updating to Remain the Same: Habitual New Media* (Cambridge, MA: MIT Press, 2016), 3; "n(you) media" appears on same page.

36 ▸ Benjamin H. Bratton, *The Stack: On Software and Sovereignty* (Cambridge, MA: MIT Press, 2016), 148.

37 ▸ Bratton, *Stack*, 163.

38 ▸ Deleuze and Guattari, *Anti-Oedipus*, 41.

39 ▸ According to Deleuze's famous "Postscript on the Societies of Control," control in contemporary societies consists primarily of codes that regulate access in "continuous variation" with the structures they encounter. Control functions in large part by addressing new scales: The individual disappears as such, to be replaced with masses, markets, and data at one end of the scalar spectrum and "dividuals," or constitutive and modulatable elements, on the other. See Gilles Deleuze, "Postscript on the Societies of Control," *October*, no. 59 (1992), 5.

40 ▸ Tim Moynihan, "Infinite Zoom Lens: How the Opening Scene of 'Limitless' Was Created" (interview), *TechHive*, March 28, 2011, https://www.techhive.com/article/223108/limitless_infinite_zoom.html.

41 ▸ Ian Failes, "Fractal Zooms and Other Side Effects in Limitless," *Fxguide* (blog), March 29, 2011, https://www.fxguide.com/featured/fractal-zooms-and-other-side-effects-in-limitless-2/.

42 ▸ And our hero, Eddie? Does he resist? Does he utilize his newfound exteriority, database access, and trans-scalar perspective to fight against the systems he now understands? Of course not. He is the perfect neoliberal subject. At the end of the film, he's running for Congress.

43 ▸ Wendy Hui Kyong Chun, *Programmed Visions: Software and Memory* (Cambridge, MA: MIT Press, 2013), 133.

44 ▸ Chun, *Programmed Visions*, 133.

45 ▸ Maier, *Theory of Relational Databases*, 3.

46 ▸ Boeke, *Cosmic View*, 4.

47 ▸ Fuller, *Media Ecologies*, 156.

48 ▸ Steve Dietz, "The Database Imaginary: Memory_Archive_Database V 4.0," in *Database Aesthetics: Art in the Age of Information Overflow* (Minneapolis: University of Minnesota Press, 2007), 114.

49 ▸ Sandra Robinson, "Databases and Doppelgängers: New Articulations of Power," *Configurations* 26, no. 4 (2018): 418, accessed November 26, 2019, https://doi.org /10.1353/con.2018.0035.

50 ▸ Lisa Nakamura, "The Socioalgorithmics of Race: Sorting It Out in Jihad Worlds," in *The New Media of Surveillance*, ed. Shoshana Magnet and Kelly Gates (Hoboken, NJ: Taylor and Francis, 2013), 150.

51 ▸ Nakamura, "Socioalgorithmics," 153.

52 ▸ James O. Coplien, "To Iterate Is Human, to Recurse, Divine," *C++ Report* 10, no. 7 (1998): 43.

53 ▸ Yuk Hui, *Recursivity and Contingency* (London: Rowman & Littlefield, 2019), 6.

54 ▸ Bratton, *Stack*, 191, 197.

55 ▸ For the classic treatment of interpellation, see Louis Althusser, "Ideology and Ideological State Apparatuses (Notes towards an Investigation)," in *Lenin and Philosophy and Other Essays* (New York: Monthly Review Press, 2001).

56 ▸ These loops form the central motif of HBO's 2016 adaptation of *Westworld* into the television series of the same name.

57 ▸ Wilfried Hou Je Bek, "Loop," in *Software Studies: A Lexicon*, ed. Matthew Fuller (Cambridge, MA: MIT Press, 2008), 182.

58 ▸ Luc Besson, dir., *Lucy*, 2014.

59 ▸ Professor Norman, during a lecture early in the film, ponders the achievements of the human race—represented through a filmic montage that spans Neolithic fire-starting (the first fire!) to the Great Pyramids, to a nuclear explosion—and asks his students, "Can we therefore conclude that humans are concerned more with having than being?"

60 ▸ The sexual import in this final segment of *Lucy*'s cosmic zoom is unmistakable, and clearly rendered in a procreative mode. Its imagery evokes, more than any narrative event, the generalized figure of life saturating all scales, from the grandest down to at least the cell. The wormhole, as transdimensional reproductive organ, bridges scalar domains at the same time it exceeds any possible anthropometric knowledge. This last sequence of images qualifies *Lucy* as a protean interdimensional zoom. The Marvel Studios extravaganza, *Doctor Strange* (dir. Scott Derrickson, 2016), contains another interdimensional zoom featuring a similar wormhole, multiple fractals and other recursive forms, a strange attractor, and various kaleidoscopic light-bending tricks. This sequence borrows variously from the legacy of the cosmic zoom, Mandelbrot's fractals, and the concept of astral projection. Interdimensionality, however, is perhaps the one subject that lies beyond the scope of this book.

61 ▸ Sherry Turkle, *Life on the Screen: Identity in the Age of the Internet* (New York: Simon & Schuster, 1997), 14.

62 ▸ Mark Poster, "Databases as Discourse; or, Electronic Interpellations," in *Computers, Surveillance, and Privacy*, ed. David Lyon and Elia Zureik (Minneapolis: University of Minnesota Press, 1996), 190.

63 ▸ Hansen, *Feed-Forward*, 6.

64 ▸ Hui, *Recursivity and Contingency*, 188.

65 ▸ Hou Je Bek, "Loop," 183.

66 ▸ Douglas Adams, *The Hitchhiker's Guide to the Galaxy* (New York: Del Rey, 1995), 21.

1 ▸ Phylis Morrison, "Wild Mice and Cosmic Views," *Sciences* 19, no. 4 (1979): 10, https://doi.org/10.1002/j.2326-1951.1979.tb01725.x.

2 ▸ Heidegger, "Age of the World Picture." 134.

3 ▸ One effect of this myopia is that both popular and academic discourse tends to analyze neoliberalism via its effects on the *individual*, rendering its true scalar functions opaque or distorted. Similarly, fascism is too often conceived of as something of a personality cult. While fascism does cathect energies on outsize public figures, this is only one node in its noxious trans-scalar networks. Humanist discourse often sets itself against the injustices wreaked by these forms of power, but ironically, humanism's scalar biases prevent a full reckoning with its purported enemies, enabling what I've called "scalar magic," where long-extant scalar forces, destructive or generative, work mostly out of sight/scale, resulting in radical surprises in the form of "sudden conjurings" that are actually no more than scalar shifts into view, such as Germany's 1939 invasion of Poland, the creation of the US Environmental Protection Agency, the 2008 financial crisis, the 2016 US presidential election, the #MeToo movement, or the COVID-19 pandemic.

4 ▸ The *USS Lexington* was scuttled after the Battle of the Coral Sea during World War II, once again a touchpoint for Boeke. Paul Allen, billionaire cofounder of Microsoft, famously located the wreck in 2018.

5 ▸ Kees Boeke, letter to James Walsh, February 28, 1958, International Institute of Social History, Amsterdam.

6 ▸ Kees Boeke, letter to James Walsh, December 8, 1960, International Institute of Social History, Amsterdam.

7 ▸ Hyman Kavett, letter to Kees Boeke, February 22, 1961, International Institute of Social History, Amsterdam.

8 ▸ Photos of Kees are liberally posted around the school and often placed within other photos taken on the school grounds, prompting a popular meme, "Waar is Kees?"— a strangely apropos example, with regard to chapter 6 of this volume, of scalar recursion.

bibliography

Adams, Douglas. *The Hitchhiker's Guide to the Galaxy*. New York: Del Rey, 1995.

Alaimo, Stacy. *Bodily Natures: Science, Environment, and the Material Self*. Bloomington: Indiana University Press, 2010.

Aldiss, Brian. "Super-Toys Last All Summer Long." *Harper's Bazaar*, December 1969.

Allen, Clile C. "Optical objective." US patent 696788 A, filed February 25, 1901, issued April 1, 1902. http://www.google.com/patents/US696788

Althusser, Louis. "Ideology and Ideological State Apparatuses (Notes towards an Investigation)." In *Lenin and Philosophy and Other Essays*, 85–126. New York: Monthly Review Press, 2001.

Amman, Othmar H., Theodore von Kármán, and Glenn B. Woodruff. "The Failure of the Tacoma Narrows Bridge." Report to the Federal Works Agency, 1941. http://resolver.caltech.edu/CaltechAUTHORS:20140512-105559175

Archimedes. *The Works of Archimedes*. Translated by Sir Thomas Heath. Mineola, NY: Dover, 2002.

Aristotle. *The Metaphysics*. London: Penguin Classics, 1999.

Barad, Karen. *Meeting the Universe Halfway: Quantum Physics and the Entanglement of Matter and Meaning*. Durham, NC: Duke University Press, 2007.

Barthes, Roland. *Camera Lucida: Reflections on Photography*. Translated by Richard Howard. New York: Hill and Wang, 2010.

Baudrillard, Jean. "The Precession of Simulacra." In *Simulacra and Simulation*, translated by Sheila Faria Glaser, 1–42. Ann Arbor: University of Michigan Press, 1994.

Bennett, Jane. *Vibrant Matter: A Political Ecology of Things*. Durham, NC: Duke University Press, 2010.

Bissonnette, Sylvie. "Scalar Travel Documentaries: Animating the Limits of the Body and Life." *Animation* 9, no. 2 (2014): 138–58. https://doi.org/10.1177/1746847714526664

Boeke, Kees. "Bilthoven, Holland's International Children's Community." *Clearing House* 13, no. 2 (1938): 106–8.

———. *Cosmic View: The Universe in Forty Jumps*. New York: John Day Co., 1957.

———. Letters to James Walsh, July 5, 1956; June 21, 1957; February 28, 1958; December 8, 1960. International Institute of Social History, Amsterdam.

———. "Sociocracy." Accessed August 31, 2014. http://worldteacher.faithweb.com /sociocracy.htm

Bohm, David. *Wholeness and the Implicate Order*. London: Routledge, 2002.

Bonner, John Tyler. *Why Size Matters: From Bacteria to Blue Whales*. Princeton, NJ: Princeton University Press, 2011.

Born, Max, and Albert Einstein. *The Born-Einstein Letters: Friendship, Politics and Physics in Uncertain Times*. Houndmills, UK: Palgrave Macmillan, 2005.

Bowker, Geoffrey C., and Susan Leigh Star. *Sorting Things Out: Classification and Its Consequences*. Rev. ed. Cambridge, MA: MIT Press, 2000.

Boyd, Danah, and Kate Crawford. "Critical Questions for Big Data." *Information, Communication & Society* 15, no. 5 (2012): 662–79. https://doi.org/10.1080/1369118X.2012 .678878

Brand, Stewart, ed. *Whole Earth Catalog*. Vol. 1, 1968.

Bratton, Benjamin H. *The Stack: On Software and Sovereignty*. Cambridge, MA: MIT Press, 2016.

Braudel, Fernand. *The Mediterranean and the Mediterranean World in the Age of Philip II*. Vol. 1. Berkeley: University of California Press, 1996.

Bryant, Levi R. *The Democracy of Objects*. Ann Arbor: MPublishing, University of Michigan Library, 2011.

Calvino, Italo. "All at One Point." In *Cosmicomics*, translated by William Weaver, 41–47. New York: Harcourt Brace Jovanovich, 1976.

Carroll, Lewis. *The Annotated Alice: The Definitive Edition*. Edited by Martin Gardner. New York: Norton, 1999.

Cheney-Lippold, John. *We Are Data: Algorithms and the Making of Our Digital Selves*. New York: NYU Press, 2017.

Chun, Wendy Hui Kyong. *Programmed Visions: Software and Memory*. Cambridge, MA: MIT Press, 2013.

———. *Updating to Remain the Same: Habitual New Media*. Cambridge, MA: MIT Press, 2016.

Cicero, Marcus Tullius. *The Dream of Scipio (De Re Publica VI 9–29)*. New York: Schwartz, Kirwin & Fauss, 1915.

Clark, Timothy. "Derangements of Scale." In *Telemorphosis: Theory in the Era of Climate Change*, edited by Tom Cohen, 1:148–66. Ann Arbor, MI: Open Humanities Press, 2012.

Codd, E. F. "A Relational Model of Data for Large Shared Data Banks." *Communications of the ACM* 13, no. 6 (1970): 377–87. https://doi.org/10.1145/362384.362685

Cohn, Jason, and Bill Jersey, directors. *Eames: The Architect & the Painter*. Documentary film. 2012.

Collins, Harry M. *Changing Order: Replication and Induction in Scientific Practice*. Chicago: University of Chicago Press, 1992.

Collins, Harry, and Trevor Pinch. *The Golem: What Everyone Should Know about Science*. Cambridge: Cambridge University Press, 1993.

Coplien, James O. "To Iterate Is Human, to Recurse, Divine." *C++ Report* 10, no. 7 (1998).

Cosgrove, Denis. *Apollo's Eye: A Cartographic Genealogy of the Earth in the Western Imagination*. Baltimore: Johns Hopkins University Press, 2003.

Date, C. J. *An Introduction to Database Systems*. Reading, MA: Addison-Wesley, 1975.

Dayan, Daniel. "The Tutor-Code of Classical Cinema." *Film Quarterly* 28, no. 1 (Autumn 1974): 22–31.

De Landa, Manuel. *A Thousand Years of Nonlinear History*. New York: Zone, 2000.

Deleuze, Gilles. *Difference and Repetition*. Translated by Paul Patton. New York: Columbia University Press, 1995.

——. "Postscript on the Societies of Control." *October*, no. 59 (1992), 3–7.

Deleuze, Gilles, and Félix Guattari. *Anti-Oedipus: Capitalism and Schizophrenia*. New York: Penguin Classics, 2009.

——. *A Thousand Plateaus: Capitalism and Schizophrenia*. Minneapolis: University of Minnesota Press, 1987.

Demetrios, Eames. *An Eames Primer*. New York: Universe, 2002.

——. *Scale Is the New Geography*. DVD, 2008.

——. "'Stations through Time' in Digital Anthology, Powers of Ten Interactive CD-ROM." Pyramid Media, 1999.

Derrida, Jacques. "Différance." In *Speech and Phenomena and Other Essays on Husserl's Theory of Signs*, translated by David B. Allison, 129–60. Evanston, IL: Northwestern University Press, 1973.

——. *Of Grammatology*. Baltimore: Johns Hopkins University Press, 1998.

——. "Structure, Sign, and Play in the Discourse of the Human Sciences." In *Writing and Difference*, translated by Alan Bass, 351–70. Chicago: University of Chicago Press, 1980.

Dietz, Steve. "The Database Imaginary: Memory_Archive_Database V 4.0." In *Database Aesthetics: Art in the Age of Information Overflow*, 110–20. Minneapolis: University of Minnesota Press, 2007.

Drexler, K. Eric. *Radical Abundance: How a Revolution in Nanotechnology Will Change Civilization*. New York: BBS PublicAffairs, 2013.

Eames, Charles, and Ray Eames. *2^n: A Story of the Power of Numbers*. Short film. 1961.

——. *A Communications Primer*. Short film. 1953.

——. *A Rough Sketch for a Proposed Film Dealing with the Powers of Ten and the Relative Size of Things in the Universe*. Short film. 1968.

——. *Blacktop: A Story of the Washing of a School Play Yard*. Short film. 1952.

——. *Day of the Dead*. Short film. 1957.

——. *Design Q&A*. Short film. 1972.

——. *Glimpses of the USA*. Multiscreen short film. 1959.

——. "Isaac Newton: Physics for a Moving Earth." Exhibition, IBM Corporate Exhibit Center, New York City, December 20, 1973.

——. "Mathematica: A World of Numbers . . . and Beyond." Exhibition, California Museum of Science and Industry, Los Angeles, March 1961.

——. *Powers of Ten: A Flipbook*. New York: W. H. Freeman, 1998.

——. *Think*. Short film. 1964.

——. *Toccata for Toy Trains*. Short film. 1957.

——. *Topology*. Short film. 1961.

——. *Tops*. Short film. 1969.

Eames Collection. Library of Congress, Prints and Photographs Division, Washington, DC.

Eames Office and DATT Japan. *Powers of Ten Interactive*. Macintosh/Windows application. Pyramid Media, 1999.

Eliot, T. S. *The Waste Land and Other Writings*. New York: Modern Library, 2002.

Elton, Sam. *A New Model of the Solar System*. New York: Philosophical Library, 1966.

Failes, Ian. "Fractal Zooms and Other Side Effects in Limitless." *Fxguide* (blog), March 29, 2011. https://www.fxguide.com/featured/fractal-zooms-and-other-side-effects-in -limitless-2/.

Foucault, Michel. *Discipline and Punish: The Birth of the Prison*. New York: Vintage, 1995.

——. *History of Madness*. Edited by Jean Khalfa. Translated by Jonathan Murphy. New York: Routledge, 2006.

——. *The Birth of the Clinic: An Archaeology of Medical Perception*. New York: Vintage, 1994.

——. *The Order of Things: An Archaeology of the Human Sciences*. New York: Vintage, 1973.

Franklin, Seb. *Control: Digitality as Cultural Logic*. Cambridge, MA: MIT Press, 2015.

French, A. P. "In Memoriam." *Physics in Perspective* 10, no. 1 (2008): 110–22. https://doi .org/10.1007/s00016-007-0343-5

Fuller, Matthew. *Media Ecologies: Materialist Energies in Art and Technoculture*. Cambridge, MA: MIT Press, 2005.

Galloway, Alexander R. *The Interface Effect*. Cambridge: Polity, 2012.

Galloway, Alexander R., and Eugene Thacker. *The Exploit: A Theory of Networks*. Minneapolis: University of Minnesota Press, 2007.

Gibson, James J. *The Ecological Approach to Visual Perception*. New York: Psychology Press, 1986.

Gimzewski, Jim, and Victoria Vesna, "The Nanomeme Syndrome: Blurring of Fact and Fiction in the Construction of a New Science." *Technoetic Arts: A Journal of Speculative Research* 1, no. 1 (January 2003): 7–24.

Ginzburg, Carlo. "Microhistory: Two or Three Things That I Know about It." Translated by John Tedeschi and Anne C. Tedeschi. *Critical Inquiry* 20, no. 1 (October 1993): 10–35.

Golec, Michael J. "Optical Constancy, Discontinuity, and Nondiscontinuity in the Eameses' Rough Sketch." In *The Educated Eye: Visual Culture and Pedagogy in the Life Sciences*, edited by Nancy Anderson and Michael R. Dietrich. Hanover, NH: Dartmouth College Press, 2012.

Golkin, Peter. "IMAX Audiences Embark on a Cosmic Voyage through Time and Space." National Air and Space Museum, June 24, 1996. http://airandspace.si.edu/about /newsroom/release/?id=111

Guattari, Felix. *Chaosmosis: An Ethico-Aesthetic Paradigm*. Translated by Julian Pefanis. Bloomington: Indiana University Press, 1995.

Hacking, Ian. *Representing and Intervening*. Cambridge: Cambridge University Press, 1983.

Haeckel, Ernst. *Generelle Morphologie der Organisme*. 2 vols. Berlin: Reimer, 1866.

Hall, Steven. *The Raw Shark Texts: A Novel*. New York: Canongate U.S., 2008.

Hansen, Mark B. N. *Feed-Forward: On the Future of Twenty-First-Century Media*. Chicago: University of Chicago Press, 2015.

Harvey, David. *The New Imperialism*. Oxford: Oxford University Press, 2005.

Harwood, Graham. "Pixel." In *Software Studies: A Lexicon*, edited by Matthew Fuller, 213–17. Cambridge, MA: MIT Press, 2008.

Hawking, Stephen. *A Brief History of Time*. 10th anniversary ed. New York: Bantam, 1998.

Hayles, N. Katherine. *How We Think: Digital Media and Contemporary Technogenesis*. Chicago: University of Chicago Press, 2012.

——. *My Mother Was a Computer: Digital Subjects and Literary Texts*. Chicago: University of Chicago Press, 2005.

Hecht, Gabrielle. "Interscalar Vehicles for an African Anthropocene: On Waste, Temporality, and Violence." *Cultural Anthropology* 33, no. 1 (2018): 109–41. https://doi.org/10.14506 /ca33.1.05

Heidegger, Martin. "The Age of the World Picture." In *The Question Concerning Technology, and Other Essays*, 115–54. New York: Harper Torchbooks, 1977.

Heise, Ursula K. *Imagining Extinction: The Cultural Meanings of Endangered Species*. Chicago: University of Chicago Press, 2016.

——. *Sense of Place and Sense of Planet: The Environmental Imagination of the Global*. New York: Oxford University Press, 2008.

Hooft, Gerard 't, and Stefan Vandoren. *Time in Powers of Ten: Natural Phenomena and Their Timescales*. Hackensack, NY: World Scientific Publishing, 2014.

Horton, Zachary. "Collapsing Scale: Nanotechnology and Geoengineering as Speculative Media." In *Shaping Emerging Technologies: Governance, Innovation, Discourse*, edited by Kornelia Konrad, Christopher Coenen, Anne Dijkstra, Colin Milburn, and Harro van Lente, 203–18. Studies of New and Emerging Technologies 4. Berlin: IOS Press, 2013.

——. "Composing a Cosmic View: Three Alternatives for Thinking Scale in the Anthropocene." In *Scale in Literature and Culture*, edited by Michael Travel Clarke and David Wittenberg, 35–60. New York: Palgrave Macmillan, 2017.

——. "The Trans-Scalar Challenge of Ecology." *Interdisciplinary Studies in Literature and Environment* 26, no. 1 (Winter 2019): 5–26.

——. "Toward a Speculative Nano-Ecology: Trans-Scalar Knowledge, Disciplinary Boundaries, and Ecology's Posthuman Horizon." *Resilience* 2, no. 3 (Fall 2015): 58–86.

Hou Je Bek, Wilfried. "Loop." In *Software Studies: A Lexicon*, edited by Matthew Fuller, 179–83. Cambridge, MA: MIT Press, 2008.

Huang, Michael, and Cary Huang. *The Scale of the Universe 2*. 2012. http://htwins.net/scale2/.

Hui, Yuk. *Recursivity and Contingency*. London: Rowman & Littlefield, 2019.

Hyder, James T. "The Making of Cosmic Voyage." edited by Maureen Kerr, Jo Hinkel, and Helen Morrill. Educational Services Department, National Air and Space Museum, Smithsonian Institution, 1996.

IBM Knowledge Center. "History of IMS: Beginnings at NASA." Accessed August 23, 2018. https://www.ibm.com/support/knowledgecenter/zosbasics/com.ibm.imsintro.doc.intro/ip0ind0011003710.htm

——. "IMS Is Strategic for Addressing Customer Needs." Accessed August 23, 2018. https://www.ibm.com/support/knowledgecenter/zosbasics/com.ibm.imsintro.doc.intro/ip0inicmst03.htm

Jameson, Fredric. *Postmodernism; or, The Cultural Logic of Late Capitalism*. Durham, NC: Duke University Press, 1999.

Jockers, Matthew L. *Macroanalysis: Digital Methods and Literary History*. Urbana: University of Illinois Press, 2013.

Joseph, Fiona. *Beatrice: The Cadbury Heiress Who Gave Away Her Fortune*. Birmingham: Foxwell Press, 2012.

Kant, Immanuel. *Critique of Judgment*. Translated by Werner S. Pluhar. Indianapolis, IN: Hackett Publishing, 1987.

Kavett, Hyman. Letter to Kees Boeke, February 22, 1961. International Institute of Social History, Amsterdam.

Kember, Sarah, and Joanna Zylinska. *Life after New Media: Mediation as a Vital Process*. Cambridge, MA: MIT Press, 2012.

Kessler, Elizabeth A. "Jumps in Space: The Disappearing and Appearing Logarithmic Scale." Society for Literature, Science, and the Arts, Dallas, TX, 2014.

——. *Picturing the Cosmos: Hubble Space Telescope Images and the Astronomical Sublime*. Minneapolis: University of Minnesota Press, 2012.

Kirk, Andrew G. *Counterculture Green: The Whole Earth Catalog and American Environmentalism*. Lawrence: University Press of Kansas, 2007.

Kirkham, Pat. *Charles and Ray Eames: Designers of the Twentieth Century*. Cambridge, MA: MIT Press, 1995.

Kowalski, Robert. "Algorithm = Logic + Control." *Communications of the ACM* 22, no. 7 (1979): 424–36.

Latour, Bruno. "Anti-Zoom." In *Scale in Literature and Culture*, edited by Michael Travel Clarke and David Wittenberg, 93–101. New York: Palgrave Macmillan, 2017.

——. *Reassembling the Social: An Introduction to Actor-Network-Theory*. New York: Oxford University Press, 2007.

<figure>256</figure>

——. "Visualisation and Cognition: Drawings Things Together." *Knowledge and Society: Studies in the Sociology of Culture Past and Present* 6 (1986): 1–40.

Lefebvre, Henri, and Donald Nicholson-Smith. *The Production of Space*. Malden, MA: Blackwell, 2011.

Locke, John. *An Essay Concerning Human Understanding*. Translated by P. H. Nidditch. Oxford: Clarendon Press, 1975.

Lucretius. *The Nature of Things*. Translated by Alicia Stallings. London: Penguin Classics, 2007.

Maier, David. *The Theory of Relational Databases*. Computer Science Press, 1983.

Mandelbrot, Benoit B. *The Fractal Geometry of Nature*. W. H. Freeman, 1982.

Manovich, Lev. *The Language of New Media*. Cambridge, MA: MIT Press, 2002.

McDaniel, Jay. "Physical Matter as Creative and Sentient." Edited by Eugene C. Hargrove. *Environmental Ethics* 5, no. 4 (1983): 291–317. https://doi.org/10.5840/enviroethics 19835413

McLuhan, Marshall. *Understanding Media: The Extensions of Man*. New York: McGraw-Hill, 1966.

Moretti, Franco. *Graphs, Maps, Trees: Abstract Models for Literary History*. New York: Verso, 2007.

Morrison, Philip. "The Actuary of Our Species." In *Nothing Is Too Wonderful to Be True*, 70–87. Woodbury, NY: American Institute of Physics, 1997.

——. "Cause, Chance, and Creation." In *Nothing Is Too Wonderful to Be True*, 88–98. Woodbury, NY: American Institute of Physics, 1997.

——. "If the Bomb Gets Out of Hand." In *Nothing Is Too Wonderful to Be True*, 291–99. Woodbury, NY: American Institute of Physics, 1997.

——. "Knowing Where You Are." In *Nothing Is Too Wonderful to Be True*, 282–90. Woodbury, NY: American Institute of Physics, 1997.

——. "Life beyond Earth and the Mind of Man." In *Nothing Is Too Wonderful to Be True*, 175–78. Woodbury, NY: American Institute of Physics, 1997.

——. "Looking at the World: An Essay." In Philip Morrison, Phylis Morrison, and Ray Eames, *Powers of Ten: About the Relative Size of Things in the Universe*, 1–15. Redding, CT: Scientific American Library, 1982.

——. "Something Personal" section introduction. In *Nothing Is Too Wonderful to Be True*, 2. Woodbury, NY: American Institute of Physics, 1997.

——. "A Talk with Philip Morrison." In *Nothing Is Too Wonderful to Be True*, 196–203. Woodbury, NY: American Institute of Physics, 1997.

——. "The Wonder of Time." In *Nothing Is Too Wonderful to Be True*, 18–20. Woodbury, NY: American Institute of Physics, 1997.

Morrison, Philip, and Phylis Morrison. "A Happy Octopus." In *The Work of Charles and Ray Eames: A Legacy of Invention*, edited by Donald Albrecht, 105–17. New York: Harry N. Abrams, 1997.

——. Preface. In Philip Morrison, Phylis Morrison, and Ray Eames, *Powers of Ten: About the Relative Size of Things in the Universe*. Redding, CT: Scientific American Library, 1982.

Morrison, Philip, Phylis Morrison, and Ray Eames. *Powers of Ten: About the Relative Size of Things in the Universe*. Redding, CT: Scientific American Library, 1982.

Morrison, Phylis. "Wild Mice and Cosmic Views." *Sciences* 19, no. 4 (1979): 10–11. https://doi.org/10.1002/j.2326-1951.1979.tb01725.x

Morton, Timothy. *Hyperobjects: Philosophy and Ecology after the End of the World*. Minneapolis: University of Minnesota Press, 2013.

——. "Thinking Ecology: The Mesh, the Strange Stranger, and the Beautiful Soul." *Collapse* 6 (2010): 265–93.

Moynihan, Tim. "Infinite Zoom Lens: How the Opening Scene of 'Limitless' Was Created" (interview). *TechHive*, March 28, 2011. https://www.techhive.com/article/223108/limit less_infinite_zoom.html

Mulvey, Laura. "Visual Pleasure and Narrative Cinema." *Screen* 16, no. 3 (1975): 6–18. https://doi.org/10.1093/screen/16.3.6

Nakamura, Lisa. "The Socioalgorithmics of Race: Sorting It Out in Jihad Worlds." In *The New Media of Surveillance*, edited by Shoshana Magnet and Kelly Gates, 149–61. Hoboken, NJ: Taylor and Francis, 2013.

Neuhart, John, Marilyn Neuhart, and Ray Eames. *Eames Design*. New York: Harry N. Abrams, 1989.

Newton, Isaac. *A Historical Account of Two Notable Corruptions of Scripture*. Seaside,OR: Rough Draft Printing, 2011.

———. *Observations upon the Prophecies of Daniel and the Apocalypse of St. John*. Gear-hart, OR: Watchmaker Publishing, 2011.

———. *The Principia: Mathematical Principles of Natural Philosophy*. Translated by Andrew Motte. 3rd ed. CreateSpace, 2013.

Nieland, Justus. "Midcentury Futurisms: Expanded Cinema, Design, and the Modernist Sensorium." *Affirmations of the Modern* 2, no. 1 (2015). http://am.ubiquitypress.com /articles/59/.

Nietzsche, Friedrich Wilhelm. *Beyond Good and Evil*. Translated by Marion Faber. Oxford: Oxford University Press, 2008.

———. "On Truth and Lie in an Extra-Moral Sense." In *The Portable Nietzsche*, translated by Walter Kaufmann, 42–47. New York: Viking, 1954.

Nixon, Rob. *Slow Violence and the Environmentalism of the Poor*. Cambridge, MA: Harvard University Press, 2011.

Paracelsus. "Hermetic Astronomy." In *The Hermetic and Alchemical Writings of Paracelsus*, edited by Arthur Edward Waite, 2:282–314. Eastford, CT: Martino Fine Books, 2009.

Parikka, Jussi. *A Geology of Media*. Minneapolis: University of Minnesota Press, 2015.

Parks, Lisa. *Cultures in Orbit: Satellites and the Televisual*. Durham, NC: Duke University Press, 2005.

Peters, John Durham. *The Marvelous Clouds: Toward a Philosophy of Elemental Media*. Chicago: University of Chicago Press, 2015.

Pickering, Andrew. *The Mangle of Practice: Time, Agency, and Science*. Chicago: University of Chicago Press, 1995.

Plato. *The Republic*. Translated by Desmond Lee. 2nd ed. London: Penguin Classics, 2003.

———. "Timaeus." In *Plato: Complete Works*, edited by John M Cooper and D. S. Hutchinson, translated by Donald J. Zeyl, 1224–91. Indianapolis: Hackett, 1997.

Plumwood, Val. *Feminism and the Mastery of Nature*. London: Routledge, 1994.

Poole, Robert. *Earthrise: How Man First Saw the Earth*. New Haven, CT: Yale University Press, 2008.

Poster, Mark. "Databases as Discourse; or, Electronic Interpellations." In *Computers, Surveillance, and Privacy*, edited by David Lyon and Elia Zureik, 175–92. Minneapolis: University of Minnesota Press, 1996.

Pryor, William T. "Maps for Engineering and Associated Work by Stages." In *Map Uses, Scales, and Accuracies for Engineering and Associated Purposes: A Report of the ASCE Surveying and Mapping Division, Committee on Cartographic Surveying*, 9–19. New York: American Society of Civil Engineers, 1983.

———. "Selection of Maps for Engineering and Associated Work." In *Map Uses, Scales, and Accuracies for Engineering and Associated Purposes: A Report of the ASCE Surveying and Mapping Division, Committee on Cartographic Surveying*, 20–62. New York: American Society of Civil Engineers, 1983.

Ratner, Mark A., and Daniel Ratner. *Nanotechnology: A Gentle Introduction to the Next Big Idea*. Upper Saddle River, NJ: Prentice Hall, 2003.

Robinson, Sandra. "Databases and Doppelgängers: New Articulations of Power." *Configura-tions* 26, no. 4 (2018): 411–40. https://doi.org/10.1353/con.2018.0035

Scharf, Caleb. *The Zoomable Universe: An Epic Tour through Cosmic Scale, from Almost Everything to Nearly Nothing*. New York: Farrar, Straus and Giroux, 2017.

Schrader, Paul. "Poetry of Ideas: The Films of Charles Eames." *Film Quarterly* 23, no. 3 (1970): 2–19.

Schuldenfrei, Eric. *The Films of Charles and Ray Eames: A Universal Sense of Expectation*. New York: Routledge, 2015.

Schweickart, Russell. "No Frames, No Boundaries." In *Earth's Answer: Explorations of Planetary Culture at the Lindisfarne Conferences*, edited by Michael Katz, 2–13. Lindisfarne Books/Harper and Row, 1977.

Silleck, Bayley, director. *Cosmic Voyage*. Short film. 1996.

Simondon, Gilbert. *On the Mode of Existence of Technical Objects*. Translated by Cecile Malaspina and John Rogove. Minneapolis: University of Minnesota Press, 2017.

Sobchack, Vivian. "Meta-Morphing." Telepolis. Accessed November 26, 2019. https://www.heise.de/tp/features/Meta-Morphing-3445963.html

———, ed. *Meta-Morphing: Visual Transformation and the Culture of Quick-Change*. Minne-apolis: University of Minnesota Press, 2000.

Spinoza, Benedict de. *Ethics*. Translated by Edwin Curley. London: Penguin Classics, 2005.

Starosielski, Nicole. *The Undersea Network*. Durham, NC: Duke University Press, 2015.

Stauffer, Robert C. "Haeckel, Darwin, and Ecology." *Quarterly Review of Biology* 32, no. 2 (June 1957): 138–44.

Stewart, Susan. *On Longing: Narratives of the Miniature, the Gigantic, the Souvenir, the Collection*. Durham, NC: Duke University Press, 1992.

Thacker, Eugene. "Networks, Swarms, Multitudes." *CTHEORY*, no. a142 (May 18, 2004). http://www.ctheory.net/articles.aspx?id=422

Theophrastus. "De Causis Plantarum." In *Selections from Early Greek Philosophy*, edited by Milton C. Nahm. 4th ed. New York: Appleton-Century-Crofts, 1964.

Thompson, D'Arcy Wentworth. *On Growth and Form*. Edited by John Tyler Bonner. Abridged ed. Cambridge: Cambridge University Press, 1992.

Turkle, Sherry. *Life on the Screen: Identity in the Age of the Internet*. New York: Simon & Schuster, 1997.

Turner, Fred. *The Democratic Surround: Multimedia and American Liberalism from World War II to the Psychedelic Sixties*. Chicago: University of Chicago Press, 2013.

Uchic, Michael D., Dennis M. Dimiduk, Jeffrey N. Florando, and William D. Nix. "Sample Dimensions Influence Strength and Crystal Plasticity." *Science* 305, no. 5686 (2004): 986–89. https://doi.org/10.1126/science.1098993

Uexküll, Jakob von. "An Introduction to Umwelt." *Semiotica* 134, no. 1/4 (2001): 107–10.

Von Baeyer, Hans Christian. "Power Tool." *Sciences* 40, no. 5 (2000): 12–15. https://doi.org/10.1002/j.2326-1951.2000.tb03516.x

Walsh, James. Letters to Kees Boeke, April 11, 1957; May 16, 1957. International Institute of Social History, Amsterdam.

Whissel, Kristen. "The Digital Multitude." *Cinema Journal* 49, no. 4 (2010): 90–110. https://doi.org/10.1353/cj.2010.0005

Wiens, J. A. "Spatial Scaling in Ecology." *Functional Ecology* 3, no. 4 (1989): 385–97.

Woods, Derek. "Epistemic Things in Charles and Ray Eames's Powers of Ten." In *Scale in Literature and Culture*, edited by Michael Travel Clarke and David Wittenberg, 61–92. New York: Palgrave Macmillan, 2017.

———. "Scale Critique for the Anthropocene." *Minnesota Review*, no. 83 (January 2014), 133–42. https://doi.org/10.1215/00265667-2782327

Zielinski, Siegfried, and Timothy Druckrey. *Deep Time of the Media: Toward an Archae-ology of Hearing and Seeing by Technical Means*. Translated by Gloria Custance. Cambridge, MA: MIT Press, 2008.

index